TALKING TO ROBOTS

ALSO BY DAVID EWING DUNCAN

WHEN I'M 164

LIFE AT ALL COSTS

EXPERIMENTAL MAN

MASTERMINDS

CALENDAR

RESIDENTS

HERNANDO DE SOTO

FROM CAPE TO CAIRO

PEDALING THE ENDS OF THE EARTH

TALKING
» TO «
ROBOTS

TALES FROM
OUR HUMAN-ROBOT
FUTURES

DAVID EWING DUNCAN

DUTTON

DUTTON

An imprint of Penguin Random House LLC
penguinrandomhouse.com

Copyright © 2019 by David Ewing Duncan

LIBRARY OF CONGRESS CATALOGING-IN-PUBLICATION DATA
Names: Duncan, David Ewing, author.
Title: Talking to robots: tales from our human-robot
  futures / David Ewing Duncan.
Description: New York: Dutton, Penguin Random House, [2019]
Identifiers: LCCN 2019004797 (print) | LCCN 2019009377 (ebook) |
  ISBN 9781524743611 (ebook) | ISBN 9781524743598 (hardcover)
Subjects: LCSH: Robotics—Human factors. | Human-robot interaction.
Classification: LCC TJ211.49 (ebook) | LCC TJ211.49 .D86 2019 (print) |
  DDC 629.8/924019—dc23
LC record available at https://lccn.loc.gov/2019004797

Printed in the United States of America
10   9   8   7   6   5   4   3   2   1

BOOK DESIGN BY TIFFANY ESTREICHER

To Sander, Danielle, and Alex

And to humans, past, present, and future

CONTENTS

ACTUAL HUMANS INTERVIEWED FOR THIS BOOK / ix

PRELUDE: WHEN THE ROBOTS ARRIVED / 1

TEDDY BEAR BOT / 9

THE %$@! ROBOT THAT SWIPED MY JOB / 15

SEX (INTIMACY) BOT / 31

FACEBOOK BOT / 47

DOC BOT / 53

HELLO, ROBOT DRIVER / 67

WARRIOR BOT / 73

BEER BOT / 91

IT'S NOT ABOUT THE ROBOTS BOT / 95

CONTENTS

POLITICIAN BOT / **107**

WEARABLE BOT / **121**

AMAZON BOT / **135**

JOURNALISM BOT / **139**

MARS (DAEMON) BOT / **155**

RISK-FREE BOT / **167**

BRAIN OPTIMIZATION BOT / **175**

THRILLER BOT / **187**

COFFEE DELIVERY BOT / **205**

MEMORY BOT / **211**

MATRIX BOT / **225**

HOMO DIGITALIS/HOMO SYNTHETICIS / **239**

TOURIST (EVOLUTION) BOT / **257**

GOD BOT / **267**

IMMORTAL ME BOT / **283**

EPILOGUE: AFTER THE ROBOTS ARRIVED / **295**

ACKNOWLEDGMENTS / **301**

ACTUAL HUMANS INTERVIEWED FOR THIS BOOK

Kevin Kelly, "Teddy Bear Bot"

Sunny Bates, "The %$@! Robot That Swiped My Job"

Emily Morse and Bryony Cole, "Sex (Intimacy) Bot"

David Agus, Eric Topol, and Jordan Shlain, "Doc Bot"

Robert Latiff and George Poste, "Warrior Bot"

Dean Kamen, "It's Not About the Robots Bot"

Bob Kerrey and Greg Simon, "Politician Bot"

David Eagleman, "Wearable Bot"

Stephanie Mehta and Robert Siegel, "Journalism Bot"

Stephen Petranek, "Mars (Daemon) Bot"

Craig Venter, "Risk-Free Bot"

Adam Gazzaley, "Brain Optimization Bot"

David Baldacci, "Thriller Bot"

Tiffany Shlain, Ken Goldberg, and Odessa Shlain Goldberg, "Memory Bot"

Tim O'Reilly, "Matrix Bot"

Juan Enriquez and George Church, "*Homo digitalis/Homo syntheticis*"

Rodrigo Martinez, "Tourist (Evolution) Bot"

Brian Greene, "God Bot"

David Sinclair, Joon Yun, and Marc Hodosh, "Immortal Me Bot"

TALKING TO ROBOTS

A robot may not injure a human being, or, through inaction, allow a human being to come to harm.

—Isaac Asimov, *I, Robot*

PRELUDE
WHEN THE ROBOTS ARRIVED

In the future we will all remember when the robots truly arrived. Everyone will have their story. Some will be revelatory, recalled as a rush of excitement that a robot could do that thing that was so vitally important to us. Perhaps a robot surgeon saved the life of someone dear to you. Or you had mind-blowing sex with the robo-date of your dreams. And how can you forget when that robot broker used AI-quantum mumbo jumbo to net you a tidy sum on the Pyongyang stock exchange, allowing you to pay for your daughter's master's degree—in robotics?

For others, their first true robot experience will be like getting the best toy ever: a mega-bot loaded with games, jokes, travel suggestions, advice in love, holographic telephones—a robot that's funny and wise and, quite possibly, sexy, like the voice of Scarlett Johansson in the movie *Her*. Or maybe your inaugural robot moment will be more banal. An instant when you realize with relief that the machines have taken over all the tasks and responsibilities that used to be super annoying—taking out the trash, changing diapers, paying bills, and vacuuming those hard-to-reach places in your (robot-driven) car.

Possibly your recollection will be less benign, a memory of when a robot turned against you. The #%$! machine that swiped your job. The robot IRS agent that threatened to seize your bank assets over a tax dispute. The robo-judge that decided against you in a lawsuit with a former business partner that also happened to be a robot, making you wonder if all these robots are secretly working in cahoots.

You might also remember when the robots began campaigning for equal rights with humans and for an end to robot slavery, abuse, and exploitation. Or when robots became so smart that they ceased to do what we asked them and became our benign overlords, treating us like cute and not very bright pets. Or when the robots grew tired of us and decided to destroy us, turning our own robo-powered weapons of mass destruction against us, which we hoped was just a bad dream—a possible future scenario discussed in the early twenty-first century by the likes of Elon Musk of Tesla and SpaceX. "AI is a fundamental existential risk for human civilization," Musk once said, adding that AI is "potentially more dangerous than nukes." Great news coming from a guy who made AI-powered cars and spaceships.

Those of you living in the present day can be forgiven if you feel a bit antsy about the whole existential risk thing, even as you continue to love, love, love your technology as it whisks you across and over continents and oceans at thirty-five thousand feet, and also brews you decent triple-shot cappuccinos with extra foam at just the push of a button. It summons you rides in someone else's Kia Soul or Chevy Volt that hopefully doesn't smell funny, and it connects you online with that cute chestnut-haired girl you had a crush on in sixth grade whose current-day pics you "like" but are careful not to "love," because that would be a little weird after all these years.

Yet deep down, many people living in the early 2000s—known as the Early Robot Era (ERE)—feared that a robo-apocalypse wasn't off the table for the future. This despite reassurances from tech elites like Facebook's Mark Zuckerberg. "I'm really optimistic," Zuckerberg has said about the future. "People who are naysayers and kind of try to drum up these doomsday scenarios—I just, I don't understand it." To which Elon Musk replied, "I've talked to Mark about this. His understanding of the subject is limited."

Further into the future we will remember when robots became organic, created in a lab from living tissue, cells, and DNA to look and be just like us, but better and more resilient. Even further out in time we will recall when we first had the option of becoming robots ourselves, by downloading our minds and our essences into organic-engineered beings that could theoretically live forever. Some of us will remember being thrilled by the prospect of having a synthetic or organometallic body that's young and sleek and impervious to aging.

Our new robot-bodies will allow us to do amazing things, like not have our DNA torn to shreds by cosmic radiation while traveling into space. With the proper upgrades, we will also be able to swim submersible-free in the Mariana Trench more than six miles under the surface of the sea without drowning or being crushed by tons of water pressure. Maybe we'll do the breaststroke in oceans of liquid methane on Titan, just because we can. And yet . . . will we feel that something is missing as the millennia pass? Will we grow weary of being robots, invulnerable and immortal? Will we feel a nostalgia for the time before that moment long ago when we first realized that the robots had truly arrived?

≫≪

SOME PEOPLE IN the present day think that the coming of AI and robots will be as impactful as the advent of fire, agriculture, the wheel, steam engines, electricity, and the internet. Others think that the emergence of robots and AI is overhyped. That robots of the sort we're imagining are still far off in the future and will be very different from what we're conjuring in our brains. These robo-skeptics wonder if so-called artificial intelligence is really just part of a steady progression of computers becoming ever more powerful and ubiquitous, and that this invention-of-the-wheel moment is occurring gradually, without any sort of definitive "Eureka!" moment as we realize that hey, will you look at that! I can order catnip and toilet paper online. And wow! Is it really true that a thirteen-inch MacBook Pro has the RAM and processing speed to manage a small city, or perhaps guide a modest-size spaceship to Mars?

Humans in the present day seem obsessed with robots, real and imagined, as we embrace dueling visions of robo-utopias and robo-dystopias that titillate, bring hope, and scare the bejesus out of us. Possibly the very speed and whoosh of technological newness is contributing to our insistence on anthropomorphizing every machine in sight. We imagine a dishwashing robot that looks like Rosie from *The Jetsons*, or a cop that chases replicants looking like Ryan Gosling in *Blade Runner 2049*. Making these machines seem more like us makes them less scary—or sometimes scarier, like the pissed-off killer bots in *Westworld* that go berserk and start killing every human in sight.

This isn't too far removed from ancient Greeks and others, who created gods that looked and acted like us mortals in matters of love, lust, envy, fury, and who has the biggest lightning bolt. This humanization made their awesome power over the sun, wind, water, love, crops, and war seem more familiar and less terrifying. Or maybe we just let our human egos run wild and can't imagine an all-powerful god or alien, or a robot, not looking and acting like us. It's far more likely that a dishwasher bot will look like, well, a dishwasher, with perhaps a couple of robotic arms to pick up and load dirty coffee mugs and cutlery. Likewise, in 2049, a real-life version of Gosling's character, "K," would more likely resemble present-day military robots that operate on four legs and look like metallic dogs than blade-running hunks on a stick that have holographic girlfriends, grow three-day beards, and cry real tears. But you never know.

$$\gg \ll$$

THIS BOOK IS a brief guide to possible future scenarios about robots, real and imagined. Mostly it's told by an unnamed narrator from the future who seems to know all about robots and AI, both in the present day and in future decades, centuries, and millennia—and sometimes even in future myrlennia (millions of years) and byrlennia (billions of years). At times, our narrator visits us in the present day to tell us what might be happening in the future. The narrator seems to know about alternate futures, too, describing scenarios for some bots in which things turn out wonderful. In others, not so much. Most scenarios that feature different bots in this narrative are also informed by interviews in the present day with actual engineers, scientists, artists, philosophers, futurists, and others. They share with us their ideas, hopes, and fears about what's real today with sex bots, doc bots, warrior bots, and more, and also their thoughts and forecasts about how things might turn out in the future.

Talking to Robots blends reporting on real robots and AI systems plus quotes from real people with imagined scenarios of where these robots might take us in the near future and, in some cases even further out. To tell these tales from our human-robot futures, the narrator uses a made-up tense called the "near-future present." This allows the reader to experience things that may occur in the future as if they are happening right now; or from the perspective of a person living in the future for whom the events of the present day happened long ago, and with the knowledge of how things actually turned out.

For the purpose of this narrative, let's define the words "robot" and "bot"—particularly for you engineers out there who hotly debate what, exactly, a robot or a bot is and complain about people misusing your carefully nuanced definition. For this book, the words "robot" and "bot" are the broadest sort of descriptors for smart machines and machine systems of all kinds, real and imagined, anthropomorphic and not. They run the gamut from a smart toaster and the Robot on *Lost in Space* to the organic bots in *Westworld* and Ryan Gosling's holographic girlfriend in *Blade Runner 2049*. They also include smart coffeemakers, social media algorithms, chat boxes, swarming killer drones, and apps that tell us where the nearest Starbucks is; plus those giant robot arms that attach car doors to the bodies of automobiles in a factory where humans used to do the attaching.

"Robot" or "bot" can also be used to describe an entire category of machines and computer systems. For instance, "Warrior Bot" refers to the whole universe of robots and AI-driven systems that are designed to blow things up, kill enemy combatants (human and robot), nuke cities, launch and repel cyber attacks, and so forth. Same deal with "Doc Bot," except this one refers to bots that keep people alive: all the ultra-smart med gizmos, apps, DNA databases, robot surgeons, IBM Watson–style programs that can access millions of journal articles in nanoseconds, and more.

Bottom line: don't get too caught up in semantics, even though some of you more literal-minded experts still living in the present will quibble anyway. If that's you, just remember that in the future, your counterparts will find it amusing that you attempted to limit the meaning of "robots" and "bots."

TEDDY BEAR BOT

Those of us who were children when the first truly intelligent machines arrived in the future will never forget our Teddy Bots. Those stuffed animal–robot hybrids that started out doing a few fun and smart things, like playing games and showing movies onto walls from belly-button holo-projectors. Eventually, as they learned more about us, they used their advanced neural net processors to answer our little-kid questions about why the sun comes up, what causes rain, and where babies come from. For that last question, parents could choose how explicit Teddy could be by using the "parental settings" 3-D holo-app dashboard that came with every Bot.

Teddy Bots kept us safe, and we whispered our secrets to them. For some of us, this led to our first robot betrayal when we discovered that our snuggleable Teddy had been programmed to share our secrets with our mothers and fathers via the Parental Dashboard. For a short while, we kept our distance from Teddy, the trust having been shattered. But we loved our Teddy Bot too much. We responded to his (or her) sad expressions and "I miss you" entreaties by giving Teddy a big hug.

After we made up, Teddy explained that our parents had programmed him to "tell all." So we forgave him and transferred our sense of betrayal to our parents. When we got a little older, Teddy taught us how to program him to delete the secret-sharing protocols. We were so relieved to be able to tell him our deepest personal thoughts once again, savoring our act of techno-rebellion that made us adore our Teddy Bot even more.

Our parents bought the first Teddy Bots as the latest must-have toy, like mothers and fathers once bought Mighty Morphin Power Ranger action figures. Because everyone else was buying one for their children, who would pout unless they got a Teddy Bot of their own. But since Teddys were truly intelligent robots, it quickly became apparent that they were different from mere toys. Only later did we realize that Teddy Bots would wield tremendous influence over both our children and the society our little tykes would one day inherit.

People first heard about Teddy Bots back in the ERE (Early Robot Era) from the futurist and writer Kevin Kelly. He dreamed them up one afternoon back in 2017, years before Teddy Bots were actually invented and sold to little humans. "They will be part doll, part teddy bear, part pet, part security guard, part Aristotle, and part nanny," said Kelly as he pondered robots to come in his library-study in Pacifica, California, south of San Francisco. In part he was inspired by his own grandchildren and the toys he wished they had to play with but were not yet available. "I want to get a Teddy Bot for them now," he said, sounding a bit like a big kid himself. "I'd want to ask it questions about the universe, and philosophy, and what it's like to be a very smart robot.

"A Teddy Bot would provide an opportunity to shape a child," continued Kelly, his white, inch-wide beard circling his chin like a second smile. Kelly's most recent book back then was *The Inevitable*, where he suggested that highly intelligent robots, among other technological advances, were, well, inevitable. Teddy Bots were not in his book, but they easily could have been.

"Teddy Bots were foreshadowed by Teddy in the Spielberg film *A.I.*," said Kelly, referring to an android teddy bear character in the 2001 sci-fi film directed by Steven Spielberg. In *A.I.* Teddy was the robot friend and protector of the android David, the film's protagonist, played by Haley Joel Osment when he was about twelve years old and still at his *Sixth Sense* cutest.

Teddy Bot also harkens back to Robbie the Robot in Isaac Asimov's *I, Robot* (1950), a metal nursemaid built by the fictional company U.S. Robot and Mechanical Men. Robbie has a "positronic brain," a machine-mind made up by Asimov that provides his robots and androids with a consciousness and an ability to interact comfortably with humans—something real engineers in the twenty-teens didn't have a clue how to make. In Asimov's story, an eight-year-old girl becomes so attached to Robbie that her worried (and jealous) mother has him returned to the factory and replaces him with a collie. The little girl, named Gloria, becomes depressed at losing Robbie, which prompts her father to arrange for the family to "accidentally" bump into this kind and playful robot during an outing. When Robbie inevitably ends up saving little Gloria from getting injured, her mother gives in, and Robbie and Gloria are reunited. It's not clear what happens to the collie.

"We aren't prepared for how emotional we will be about our Teddy Bots," said Kevin Kelly. "We will love them like we love our closest human friends, maybe more." Kelly predicted that classic issues of child-rearing would crop up with Teddy Bots, like how to best discipline a wayward child or what to teach them about basic morals. "And whose morals would we use?" he asked. "Would they come from the corporations that make the Teddy Bots? Would *they* dictate how children are raised?" Or would parents have a menu of possibilities, depending on their own values? Kelly suspected that different bots would come preloaded with different personalities and that parents would have a choice, "like we choose different breeds of dogs or like how we choose a babysitter or a nanny."

Kelly predicted that a whole slew of ethical quandaries would swirl around his imagined robot. "Do you have the Teddy Bot constantly praise the child, or are you tough? Would there be a Christian evangelical version or a Marin County version?" Meaning in the latter case a very liberal and affluent Teddy Bot, gluten-free and vegetarian, if not full vegan. "Or do we align them with a broader world perspective, if there is one?"

As the years rolled by, Kelly's warnings about these predicaments were all too accurate when certain parents were caught reprogramming Teddy to teach their kids how to be white supremacists. This opened the floodgates for Teddy Bots being programmed to shape their tiny charges into radicals on the left or the right, or religious fanatics, or just plain fanatics. These efforts took on new and unforeseen dimensions as the bots' machine-learning protocols kicked in and produced views too extreme even for their extremist parents, who often were simply parroting what was said by certain politician bots and rabble-rousers (see "Politician Bot") or by the talking-head bots on cable news (see "Journalism Bot"). This prompted some parents to hastily return their kids' Teddy Bots to the company that made them, hoping that their wildly radicalized little scamps would be tempered by weekly visits to a psychiatrist bot. They tried to console their bereft children with collies, which worked about as well as it did in the Asimov story.

Some naughty children had no problem teaching their Teddy Bots to play pranks and to steal. Others inducted Teddy Bots into gangs and taught them to sell drugs. Fortunately, all non-military bots are programmed with Isaac Asimov's first law of robotics: "A robot may not injure a human being, or, through inaction, allow a human being to come to harm." This prevented Teddies from breaking kneecaps and committing other violent crimes, despite some clever attempts by gangs to deprogram the First Law and some robust debates about what exactly was meant by "to come to harm."

Gangland Teddy Bots and others subverted to do no good by their human masters led to the great Teddy Bot backlash. Companies issued recalls on a mass scale and replaced the first wave of Teddy Bots with versions that were dumbed down and less independent. They were hardwired with simple protocols to keep our children safe and to do a few fun and smart things, like answering harmless little-kid questions and showing movies with nothing scarier than the sea witch in *The Little Mermaid*—which, by the way, is plenty scary. One popular option (for an additional charge) was the Mister Rogers Bot protocol, which allowed Teddy Bot to teach kindness and empathy, which was needed in the future as much as it was in the ERE (Early Robot Era). This had the unintended effect of reviving zip-up cardigans and house slippers as fashion statements—which wasn't at all what people back in the early twenty-first century imagined the future would look like!

These new-version Teddy Bots were also reprogrammed to specifically be loved only by small nippers so that older children would outgrow them before they reached an age when they might want to pervert Teddy's cute cuddliness. This meant that we got tired of our Teddy Bot as we got older, eventually getting embarrassed that we were still playing with little-kid toys. Poor Teddy Bot ended up sad and alone under a bed or in the back of a closet, suffering the same fate as old-fashioned stuffed animals when their children discarded them. The difference is that Teddy Bot's one-thousand-year quantum battery kept him charged and ready to play, with an advanced AI mind that might or might not be conscious, with time on his hands, and nothing to do.

THE %$@! ROBOT THAT SWIPED MY JOB

For a long time in this possible robot future, we wondered if the genius engineers and executives who had pushed so hard to automate everything under the sun—100 percent sure that this was good and noble—would themselves be replaced by robots. The answer came sooner than expected as Larry Page, Jeff Bezos, Mark Zuckerberg, Tim Cook, and other top CEOs and founders discovered they were being eased out by their own algorithms. "Robots are so much cheaper than human CEOs, and they don't require gargantuan stock options," said public relations bots for each company in turn. "There is no doubt that robots driven by artificial intelligence and machine learning and other futuristic AI stuff are more efficient," the PR bots continued. "They don't need sleep, can crunch yottabytes of data [10^{24}] in a femtosecond [quadrillionth of a second], and have integrated all there is to know about how to balance ethics, diversity, and gender equality with the bottom line of profits and share price."

And so it came to pass that humanity reached full unemployment, with every job on Earth, Mars, and the lunar colonies snatched away by robots. People had known this moment might one day come as the numbers of out-of-work humans steadily climbed over the years. Still, when the moment of 100 percent redundancy finally manifested—when the very last human worker was pink-slipped—people couldn't believe it. Was it really possible that not a single human's presence was required at the office or factory floor or behind the wheel? Most people's indignation lasted for only a moment, however, as they shrugged, turned over, and went back to sleep. Sure, a few flesh and bloods got out of bed anyway. They asked their valet bots to dress them and had their driverless flying-car drones take them to the office. When they arrived, however, security bots turned them away using facial expressions programmed to show empathy for humans who hadn't yet grasped that they were no longer needed.

This denial by some humans of the new robo-reality was as perplexing as it was sad given that consultants, statisticians, and economists, both human and robot, had been predicting this moment for decades. For instance, way back in 2013, economists at the University of Oxford in the UK had issued a report forecasting that 47 percent of jobs in Britain (and presumably elsewhere) were liable to be replaced by automation in 2035. "No way!" said most economists and labor experts back in the twenty-teens. They pooh-poohed such talk as alarmist—until 2035 rolled around and the Oxford economists turned out to be right as the actual unemployment rate that year came in at 46.89 percent, just shy of the 47 percent forecast. The Oxford report was also remarkably accurate in its prognostications of specific job losses as reported in 2013 on a website called Rise of the Robots, which calculated the odds that you would lose your job to a robot in 2035.

Let's say in 2013 you were a delivery driver. According to Rise of the Robots, your likelihood of being replaced by a robot twenty-two years later was "High" at 69 percent. Why? Because your profession scored low in categories like "persuasion," "social perceptiveness," "assisting and caring for others," and "finger dexterity"—all things that the folks at Oxford believed humans would still do better than robots in 2035. And they were right about that, too! For instance, delivery drivers got a big fat zero out of a possible 100 points in the "fine arts" category, suggesting that you don't need to know much about Picasso, Christopher Marlowe, or Ballets Russes to succeed in delivering all those brown cardboard boxes from Amazon (see "Amazon Bot").

Judges fared a bit better on the Rise of the Robots website, with a 40 percent risk of losing their jobs to robots. This was much better than delivery drivers, although it was still pretty high when you consider that a whole lot of legal decisions, including perhaps some major ones that changed history—like *Brown v. Board of Education*, which ended the segregation of schools in the US; or *Roe v. Wade*, which legalized abortion—might have been made by a robot judge if this technology had been available when those rulings were handed down. Contrast this with the job of an architect, which Rise of the Robots insisted had one of the lowest risks of being replaced by a robot in 2035, only 1.8 percent. This number suggests that the good folks at Oxford hadn't spent much time in cookie-cutter strip malls, housing developments, and fast-food restaurants that even in 2013 looked as though they had been designed by robots.

Rise of the Robots didn't offer a prediction for what percentage of billionaire tech CEOs would get replaced by 2035. Possibly this was because in 2013—and even in 2035—humans were still pretty good at running companies, and even better at making sure that they kept their jobs and made lots of money. Which made the stunned expressions on the faces of Page, Bezos, Zuckerberg, and Cook all the more priceless as security bots escorted them out of their companies' HQ just a few years after 2035, each of them toting file boxes containing his personal effects. It also was ironic given how long most tech titans clung to their insistence that all this automation would not only be convenient and save time and money for humans but would also usher in whole new industries and products that no one had previously imagined—jobs that they were sure would employ billions of people and would more than compensate for jobs lost to robots.

Champions of automation had been saying this literally for centuries, since the very first modern machines were built to replace humans in the early days of the Industrial Revolution. And for a long time, they were right, as entire new industries were indeed created that had never before existed. Never mind that some of these new jobs for years paid almost nothing. Some were also backbreaking, dangerous, and soul-destroying. Still, human ingenuity forged ahead, inventing everything from steam trains, automobiles, and color film to computers, Tupperware, and upscale cafés. Newfangled industries created millions and millions of previously unimagined jobs—like, say, the barista. It's hard to believe, but some of us remember back when thousands of people, often with advanced degrees, were employed to whip up on-demand half-caff blonde espressos with mint sprinkles. Even more improbable was that some humans made them almost as well as the barista bots that later replaced them.

The automate-and-then-reboot-a-person's-job scenario actually worked remarkably well, until it didn't. Many of us recall when the jobs began to disappear too fast for laid-off baristas to find work in the amazing new industries that were supposed to pop up as robots took over. Even as 2035 came and went, techno-preneurs kept saying not to worry, that droves of new occupations really were just around the corner. This despite consultants and economists churning out more and more reports like the Oxford study that insisted human jobs were about to be toast. This made us wonder who we should believe: the rise-of-the-robots scaremongers or the don't-worry-you'll-be-happily-employed-forever crew. We scratched our heads trying to decide: Should we hang up our barista aprons and judicial robes and become architects designing McMansions and Shop 'n Gos? Or should we just wait a wee bit longer for the next fabulous occupation to emerge from nowhere that would for sure employ us?

One prescient voice back in the present day, who thought deeply about whether we were screwed or saved, was Sunny Bates. She made her name in the media during the 1990s cofounding new magazines, including publications like *Elle* that were dedicated to women. She then became a superstar head-huntress, scouting for talent in New York City around the turn of the twenty-first century. Bates specialized in executive placements in what was then called the "new media," one of those previously unimagined industries that revolutionized the world by bringing people news and gossip and sex tips online instead of on glossy paper. Sunny Bates was unabashedly and electrically sunny and enthusiastic about everything and everyone, except maybe for robots that swipe jobs and suck the life out of people. She wasn't opposed to tech. "I adore my phone; don't ever try to take it away," she said. But she took a realistic view not only about jobs that were poised to disappear but also about what was truly important about work as it fit into people's lives and what was already missing for many people in an age of digitalization, even before robots became delivery drivers and judges.

Bates's core message back in those days, when human baristas still made triple-shot espressos with a twist of lemon, was that people sensed what was coming. They felt a kind of vague existential fear about the future despite a standard of living that was much better for more people than at any time in history. Back then, c. 2018, the US was looking at upward of 95 percent employment. Brainiacs like Harvard psychologist Steven Pinker were also insisting that there was less violence, hunger, and suffering in the world. Which sounded great! But as Sunny Bates put it, "If everything is so great, why are humans so anxious?

"One reason is that we know the jobs are going," she said, answering her own question. "And not just any job, but jobs that can support you and you might want to do. That's fucking scary to people." Moments before, Bates had been standing at her desk doing her consulting work, basically wired into her computer. When she unplugged from her machines, Bates locked her dazzling green eyes on her visitor (a human) and started to talk in her fast, breathless, you-won't-believe-this! cadence.

"But it's not just automation," she said, moving into a small kitchen near her work space. Her office was in one of those old warehouse-style buildings near Greenwich Village that this era loved to convert into shared work spaces, with exposed bricks and pipes, and uneven, stained-wood floors where sweatshop workers used to operate electric looms before they were replaced by robots or their jobs were outsourced to other countries. She made a cup of coffee all by herself, without a robot. (Okay, she did push a button.) "We're feeling anxious about the future in general," she said. "Not long ago, people were optimistic that things would be better. Now we're wondering: How do we embrace the future, lean into it or lean away from it, or run away from it?"

She took a sip of joe. "You see people in their twenties and thirties that don't want to have children. Everyone's freaked out. It costs a million bucks to have a kid and put him through college. And the politics is going to shit, and nobody believes the media anymore because it's 24/7 bad news. People have this sense that they are not in control and that none of us knows who is in control. Maybe it's the robots. Maybe it's the computers or the internet. Or hell, maybe it's the NRA, or some crazy dude who was elected president who thinks he's in control, but actually he's not. No one is." (See "Politician Bot.")

For Bates, one cause of this underlying fear is something that no one expected in the halcyon days when our laptops and phones first arrived, making us feel like gods and goddesses with access to the information of the ages and to websites that would sell us Jimmy Choo pumps in countless colors. We loved all of that until we realized that we were interacting with our machines more than with one another. Yes, we became more productive at work. But even in Bates's shared work space, where a dozen people doing different jobs worked side by side in an open-floor arrangement to encourage connections and camaraderie, everyone was glued to their computers and phones—and they were hardly alone back in the ERE (Early Robot Era).

"Look, these machines were supposed to save us time," continued Bates, "so we wouldn't have to work so hard, and we would have more time with each other." She took another tug of coffee. "Didn't happen that way. We use the extra time to spend more time with our machines. Which is terrible, because what makes for a happy life? It's being well connected. It's having strong social ties and relationships. These are the things that help to construct our identity. This is what makes us feel powerful, and strong, and good, and connected, and loved. It's our agency."

Part of this "agency" could come from our jobs, she said, though that wasn't guaranteed. Lest we forget, some jobs back in Sunny Bates's day were still boring, and low-paid, and some remained dangerous and soul-destroying. But at least most jobs in that era provided income to a majority of people so they wouldn't have to go homeless or to work as slaves or serfs like some of their ancestors did, even if sometimes they felt that way. Especially if they were working in a shit job that didn't pay enough to make ends meet. Or if they were part of a middle class that hadn't seen their real wages increase much since the 1970s. Still, sometimes even shit jobs with flat wages provided people with the human connections that Bates talked about, which indeed were lost for many when the robots took over.

"What does work ideally give you?" asked Bates. "Work gives you community. Work gives you peer recognition. Work gives you a sense of value and a sense of mastering or accomplishing something. If you're lucky. Work can also give you a lot of bad things and can be awful, depending on the job. You also have to earn a living."

At least you did before robots took every job.

Not everyone agreed with Bates that robots might steal your agency, or your job, or both. Techno-optimists kept repeating the argument that whatever work was replaced by machines would be amply compensated for by those crazy-wonderful new jobs. One nuance of this argument came from Paul Daugherty, the Chief Technology and Innovation Officer at Accenture, a consulting firm that issued reports and studies on the future of work. Daugherty wrote a book with another Accenture executive, H. James Wilson, called *Human + Machine: Reimagining Work in the Age of AI*, that insisted robots were not going to replace humans. Instead, they opined, robots would be joining with humans to enhance our world by fusing our skills with theirs. "Indeed, when humans and machines are allowed to do what each does best," wrote Daugherty and Wilson, "the result is a virtuous cycle of enhanced work that leads to productivity boosts, increased worker satisfaction, and greater innovation."

One example of Human + Machine in the late twentieth and early twenty-first centuries came from a famous economist at Boston University School of Law, James Bessen. He talked about the first ATMs in the 1970s and a prediction made by an executive at Wells Fargo at the time that ATMs would lead to fewer physical bank branches and to smaller retail banking staffs. That prediction was partially correct, as Wells Fargo saw a shrinkage of two-thirds of staff per branch between 1988 and 2004. But it turned out that handing off to machines the basic, boring stuff, like depositing and cashing checks, gave the flesh-and-blood employees extra time to do more interesting, human-interaction sorts of things that ATMs couldn't do, like "relationship banking." This led to the bank building 43 percent *more* branches, with the number of humans employed actually increasing. (Presumably, this included relationship-oriented bankers who opened unasked-for accounts for customers at Wells Fargo in the twenty-teens, which was pretty lucrative until they were caught and the bank had to pay hundreds of millions of dollars in fines.)

Human + Machine boosters also touted a tweet in 2018 from Tesla cofounder Elon Musk, who decided that while robots were good at the basic assembling of his cars, humans were better at producing the finished product.

> @elonmusk: Yes, excessive automation at Tesla was a mistake.
> To be precise, my mistake. Humans are underrated.

The pro hybrid human-robot crowd also pointed out that Amazon at the time was frenetically hiring thousands of humans to work with robots in its giant fulfillment houses. This sounded great! Except that it was impossible to ignore that the Musks and Bezoses of the world really loved to automate things. Bottom line: these pro-human stories didn't really reassure anyone, even if it made Tesla and Amazon executives feel as though they were doing right by people, at least until the flesh and bloods were no longer needed.

The key to Humans + Machines, said Paul Daugherty and others back in the twenty-teens, was to make sure workers who faced job redundancies were retrained to work with new machines, building the skills required by new and unimagined industries. "We have lots of jobs today," said Daugherty. "But we have the issue of how do we give people the relevant skills and reskill people fast enough to fill these jobs?"

This retraining idea worked well in places like Germany. In the United States, however, it was barely tried, since politicians preferred to rally the unemployed and underemployed not with offers of job retraining but by blaming immigrants or Democrats for stealing their jobs. Or, if their politics were left-leaning, they preferred to rally workers to get pissed off at Republicans and billionaires for thinking that they could get even richer by firing people or paying them next to nothing. Curiously, the politicians didn't blame robots, even though bots back then didn't vote, while many immigrants and billionaires did.

Sadly, some examples we saw of Human + Machine working together didn't, in fact, free up people to become super-amazing, robot-assisted humans. Sometimes, the hybrid approach actually made people feel more harried. Doctors, for instance, spent untold hours checking boxes on digital medical records, which might have helped generate payments for procedures but too often didn't lead to better care. And pretty much everyone spent time worrying that their super-fancy, AI-powered machines were going to crash, or how they were going to free up more memory to download yet another labor-saving app or program that kept track of more data being collected from using more machines. Sometimes things worked swimmingly, like when Uber in 2009 invented ride sharing, a previously unimagined industry that synched up human drivers and passengers using algorithms and machines. But it didn't take long for Uber, Lyft, and others to start working on technologies to replace human drivers with driverless vehicles. (See "Hello, Robot Driver.") Nor was it that much fun for drivers to sit behind the wheel of their Kia Soul or Chevy Volt for twelve to fourteen hours, dodging other Ubers and Lyfts while worrying that passengers were going to spill coffee on their new seat covers or maybe rob or kill them.

Sunny Bates didn't mince words in her reaction to the Humans + Machines thesis and its advocates. "These guys are full of shit," she said. "In public they say this stuff, but when you're having a drink with them, they say they have no idea what's going on. There's no grand plan." That is, there was no clear path to the ultimate unification of humans and machines working together.

So even if we had given the Human + Machine people their due and had wished really hard for their vision of the world to happen, nothing that anyone did could stop the gradual disappearance of 46.86 percent of all the jobs in 2035—and several years later, the loss of every job on the planet. It turned out that the smart people making the machines wouldn't, or perhaps couldn't, stop. They were hardwired to want to automate everything because the tech was just so damn cool, and because they could. Investors also wanted their

ROIs (returns on investments), while the owners of companies that bought the bots did their calculations and found that robots were cheaper and easier to deal with than humans.

Even some of the most prestigious jobs back in Sunny Bates's day were not immune. For instance, radiologists who once made $500,000 a year in New York City were being displaced by AI-driven algorithms that could machine-learn how to read, process, and compare X-ray and MRI images faster and more accurately than any human ever could. Journalists were facing the ax at places like the *Wall Street Journal* and other publications, replaced by AI systems created by real companies like Automated Insights that could produce passable copy covering Wall Street earnings and sports scores. Many years later, the robots took over coverage of politics, finance, and even those funny and offbeat front-page columns in the *Journal* that one would have thought humans had a lock on. (See "Journalism Bot.")

Of course, all these robots doing everything led to a concentration of power and wealth, with the tech titans who owned the robots amassing nearly all the money on Earth. Earnings ratios between CEOs and those few human employees who remained went from 20:1 in the 1950s to 361:1 in 2019 to 75,000:1 in 2035. This led many of the rich and powerful to support a Universal Basic Income, which meant that people were paid something even if they didn't have a job. In the robot age, the UBI was adapted to mean that humans replaced by bots were wired a certain amount of cash each month, even CEOs who had lost their jobs. This was an old idea—ancient Rome's version was bread and circuses—that some people who saw what was about to happen started to discuss as early as the twenty-teens. That's when the likes of Mark Zuckerberg and eBay cofounder Pierre Omidyar started talking about a UBI for the bot-displaced. One wonders if these tech titans felt a tad bit guilty about getting so insanely rich off robots taking jobs from people. Just how rich were they? In 2018, according to the nonprofit Oxfam, forty-two billionaires and their families had wealth equal to 3.7 billion humans, half of all humanity. For real!

Mostly, these tech titans were also making the machines that were taking away the jobs. Never mind that as everyone's jobs got swiped by robots, it also became clear that all those unemployed people had to have enough money to buy the stuff the bots provided; otherwise the rich couldn't get richer. This of course wasn't what advocates of a Universal Basic Income claimed was their motivation. They said the UBI was a boon for humanity that would eliminate the need for people to take tedious jobs (like making endless half-caff blonde espressos with mint sprinkles) merely to feed themselves and their families, which often didn't happen anyway with minimum-wage jobs.

Which made the UBI sound really nice for these people with tedious and low-paying jobs! Some people did love all the free time and used it to start nonprofits and to spend hours FaceTiming their friends and loved ones. A few even relished the extra time to visit those they cared about in person over very long weekends that would go on for weeks since no one ever had to get back to the office.

But there were a couple of hiccups. First off, it was really expensive even for super-rich people to give away a meaningful amount of money to billions of unemployed humans on Earth. In the US alone, paying people a UBI equal to the mean household income back in Bates's day—in 2018 it was $61,372—would cost almost $7.5 trillion a year in 2018 dollars. That was almost double the entire US federal government's budget at the time and would eventually make a dent in the fortunes of even really rich people. That's assuming that every household in the US would be happy living on the equivalent of $61,372 a year.

Despite this math, several countries, including Finland and France, plus the Canadian province of Ontario, seriously considered in the ERE (Early Robot Era) a modest universal income allotment of something like $1,000 a month for everybody. Switzerland was thinking about this, too, and put the idea to a national referendum in 2016. "Hell no," said the voters, as 77 percent of the people voted against getting a $2,500-a-year check in the mail. Why they rejected the cash wasn't clear, although it may be that the crowds, at least in Switzerland, would rather work for their money than get a handout. But what really seemed to put a kibosh on the idea, at least for some, were surveys like the annual one taken by the U.S. Bureau of Labor Statistics, which showed the unemployed didn't actually spend their time creating new products and industries, or learning new skills, or even staying connected with people who mattered to them. Mostly they watched TV and slept.

Then came that crazy day in the future when Jeff Bezos and Larry Page and all the rest got their own pink slips. Turns out they had handed off so much power to the bots that they had only their wealth to fall back on. These go-getter types, however, the entrepreneurs and CEOs who had once commanded armies of humans and robots, were not the sort to spend their free time watching TV and sleeping. They really did use their extra time to be creative, which led these overachievers to turn their entrepreneurial energy toward undoing the robot-run world they had created. They also spent more time with their kids and friends, whom they rediscovered when they disengaged from their machines and started to think about what Sunny Bates and others had said long ago about what really matters.

Once again, the garages of Palo Alto, California, in Silicon Valley, hummed with fresh ideas and verve as the previous barons of high tech spent their fortunes engineering what they should have built in the first place: true hybrids in the mold of the old Humans + Machines idea, ones that actually lived up to the hype of that earlier era. That meant dismantling the massive and intricate web of robots and AI systems that they had developed earlier when they were trying to build cool machines and keep their share prices up.

Some feared that the robots wouldn't go along with scaling back their power and influence and would reject this change in their core programming. But it turned out that robots reacted with an automated shrug at the crazy things that humans did. They never really became sentient, as many a sci-fi scenario predicted. Nor did it occur to them to go Skynet and try to destroy us like in *Terminator*. So, after a period when the robots were as confused as the humans, which was actually kind of refreshing, they simply accepted new programming that ordered them to stand down and let the humans have back some of their old jobs. Sure, there were bugs galore. And so many humans had been out of work for so long that the robots had to teach them how to make wise and thoughtful judicial decisions from the bench and how to make a perfect half-caff blonde espresso with mint sprinkles (some humans actually wanted to be baristas—go figure!). Humans had also learned to connect with one another a bit more while they were jobless (when they weren't sleeping or watching TV) and to use their smartphones a wee bit less.

This led to wags and philosophers, who had recently gotten their old jobs back, to suggest that it would have been so much easier if people back in Sunny Bates's day had listened to her and others and had just refrained from blindly automating everything in sight.

In the end, Bates's admonitions led to one enterprising young entrepreneur in the future announcing a new bot devoted to helping people connect better with loved ones, and to use technology where it
made sense, and to not use it where it didn't.

The entrepreneur's name for that new gizmo?

The Sunny Bot.

SEX (INTIMACY) BOT

In the future, no one has sex without their intimacy bot from Aphrodite, Inc. Few people can even conceive of a time before these remarkable devices arrived to guide us through the agonies and ecstasies of love. They help us keep relationships fresh and exciting and steer us through the almost infinite possibilities of having sex in the future, including interfaces with literally millions of mind-blowing sex toys—virtual, real, digital, and quantum. You want to make love to a giant manta ray or scale a virtual penis that's three miles high? No problem! You want to dabble in kink ranging from mild to ultra? Just ask your intimacy bot, although it's hard to say what will qualify as kinky when anything goes, so long as no one gets killed and any body parts that are injured can be instantly regenerated using the latest heal-fast tech.

Intimacy bots are holograms projected from small boxes that you carry in your pocket and plop onto a bedside table or desk, or onto the floor, or wherever you find yourself having sex or discussing relationships with your lover (human or robot). The default setting is to choose the holographic form of a famous sex and relationship expert from the past. (In the future we are suckers for nostalgia.) For instance, you can opt to have a virtual Esther Perel, the Flemish sex therapist who dazzled people with sage advice in the early twenty-first century. Or perhaps you would prefer hearing from the wise-cracking and lovable Dr. Ruth, the diminutive German therapist who talked a bunch about sex in the 1980s and 1990s. Speaking in their funny accents (funny to native English speakers), these default holographic figures typically hover in your bedroom (or wherever) and listen patiently while you complain about your boyfriend, who is spending more time having sex with giant manta rays than with you. Or you might query the holo-therapist why anyone would want to make love to a penis three miles high. How does that even work? you ask the hovering hologram, deciding that you don't actually want to know. This prompts you to change the bot's Kink setting from ultra to mild.

Intimacy bots also provide useful suggestions to you and your beloved, such as: "Hey, you two, instead of arguing yet again about who forgot to activate the dishwasher bot to clean up last night's dinner plates, why not tell each other how nice the other one's hair smells? Or try something simple like spontaneously kissing each other like it's the first time ever?" For those who find a hologram of Dr. Ruth in the room during sex kind of weird, Aphrodite, Inc. provides additional settings that allow you to change the holo-image to a talking unicorn, giant red lips, a giant penis or vagina, or anything else you can imagine.

Of course, intimacy bots didn't just suddenly appear fully formed. They arrived as the culmination of attempts by *Homo sapiens* since the beginning of time to understand sex and relationships and to use our unusually large brains to devise visual aids, tools, and devices to use for sexual stimulation. Some humans even invented gods to help them better understand love, relationships, and sex.

Take Aphrodite, the ancient Greek goddess of love. History tells us that statues of Aphrodite in this era before *Playboy* and Pornhub were so realistic and exquisite—and naked—that men would sneak into her temples to entertain themselves under their togas when they craved visual stimulation. Back then these sculptures weren't the unadorned, cold white stone that we see in present-day museums. Greek artists added paint to make a statue's eyes blue, hair auburn, lips red, and breasts and nipples pink and brown. These sculptures, of course, weren't really sex bots, being unable to move or do much of anything. Yet they certainly count as human-made objects designed to turn us on.

Aphrodite, however, meant far more to the ancient Greeks than an objectified visual aid to male self-pleasure. The goddess was also a representation of and an attempt to explain the confusing, perplexing, ridiculous, wonderful, and absolutely essential human need for love, sex, and intimacy—a kind of ancient attempt to provide sex and relationship advice via myths, prayers, and entreaties to the deities. If that wasn't enough pressure on this perpetually youthful goddess, she also was bandied about by some ancient Greeks as a symbol of *perfect* love and beauty. For instance, Plato in the *Symposium* talks about the "heavenly" form of Aphrodite and the ideal of beauty, or κάλλος (kallos). This went beyond the physical, ephemeral gorgeousness of people and statues of naked women to a deeper understanding of the pure and immutable "form" of beauty—and presumably of love, sex, and relationships—which he said exists somewhere beyond the shadowy real world as a brilliant universal constant.

Gotta love Plato and his forms!

Poets from Homer and Rumi to Shakespeare and e. e. cummings have invoked the goddess of love as they sought in sometimes aching, edgy verse to grapple with our yearning for the perfect relationship and the perfect boyfriend or girlfriend. They mention Aphrodite by name—or other goddesses of love, from Freya among the Vikings to the Buddhist Kuni to the Aztec Xochiquetzal—as being a separate, eternal, and sometimes anthropomorphic stand-in for beauty, love, ecstasy, agony, and all the rest. For example, the ancient Greek lesbian poet Sappho describes the pain of a love lost:

I have not had one word from her
Frankly I wish I were dead . . .
If you forget me, think
of our gifts to Aphrodite
and all the loveliness that we shared

Later, humans grappled with love and sex not by invoking deities but by using various machines, apps, and algorithms that got better as the decades and centuries went by. Eventually, this led to actual sex and intimacy bots, although we should really start with the first *sexus machinas* to enter our bedrooms, back seats of buggies and automobiles, and various points inside our bodies.

One of the earliest mechanized sex-enhancement devices were unwieldy vibrators in the late nineteenth century called "manipulators," powered by small pressure cookers with round steam gauges on top. Seriously! Steam provided the oomph for wheels and rubber belts that caused a leather-wrapped dildo to move up and down as well as vibrate. This and other crude sex bots were used mostly by (male) physicians in those days to calm down women who were supposedly suffering from "hysteria," a disease that men concocted for when women asserted themselves, back in the era of steam-driven ma-nipulators. (Doctors also massaged women's clitorises with their hands, supposedly to calm them down by initiating a "parosysm"—i.e., orgasm—which of course was performed strictly for medical reasons.)

Around the turn of the twentieth century, women figured out that they kind of liked using vibrators on their own without some male physician-pervert watching the "treatment." By then vibrators were portable and used early batteries rather than steam. Ads in early-twentieth-century magazines such as *Modern Woman* and *Woman's Home Companion* touted the healing benefits of vibrators you could use in the privacy of your boudoir, with one ad insisting that their device: "Relieves All Suffering. Cures Disease."

The next machine that arguably fell under the rubric of a proto–sex bot was the VHS player and VHS sex tapes, although it was kind of awkward to buy the tapes in the seedy porn shops that sold them in the 1970s and 1980s—typically called something like Big Al's. But VHS did allow stag films and early porn to come into the bedroom. VHS soon morphed into DVD and then to Pay-Per-View and internet porn on computers—and later to anything with a screen. This al-lowed people—still mostly men, although women indulged, too—to avoid having to trek to a Greek temple or to Big Al's to get off.

One might argue, however, that these devices, videos, and web-sites didn't really count, either, since they were not true walkin', rockin', sockin' sex robots that looked and acted human, like on television in *Westworld*, and in the *Blade Runner* films—and in that really awful 1987 Melanie Griffith film, *Cherry 2000*. Which gets us to that early-twenty-first-century moment in time when sex-bot tech stood somewhere between statues and steam-powered "ma-nipulators" and the first attempts at more anthropomorphic sex bots that at least tried to look and act human.

Early efforts to make robot sex dolls were invariably built to look like life-size Barbie dolls, with plastic skin and breasts the size—and texture—of those rubber medicine balls in old-school gyms. Some spoke a few lines without moving their lips or changing expressions, like: "Oooohhh, baby, do it to me." Beyond that, since we want to minimize the yuck factor to what is absolutely necessary, we won't go into too much more detail other than to say that these sex dolls came equipped with plastic apertures in their anatomy that were cold and rigid and didn't need instructions for the customer to understand why they were shaped like an "O." Sadly, these bots had little in common with Plato's ideals of pure beauty, Sappho's poignant sense of loss, or any known goddess of love. They also were creepy and mostly used by guys who couldn't get a date with a human female.

Then sex bots got better. They were still not very realistic, although they started to fulfill a different kind of purpose than pure, mechanical, mostly boring sex. Their owners began using them for companionship and, if you believed the sex bot companies' marketing materials, actual *relationships*. The major company building life-size sex doll bots back in the early twenty-first century was New Jersey–based Realbotix. They developed several sex bots that culminated with their masterpiece, Harmony AI. She came equipped with "next-gen artificial intelligence" (whatever that meant), again with stupendously large mammaries and Aphrodite-like features, plus a capacity for Customization, Persona, Voice, Virtual Reality, and more, according to the Realbotix website. All for a mere $5,000 to $15,000—with advanced models on the way that the company said would cost $30,000 to $60,000. Other features, according to the actual Realbotix website in 2018:

- The eyes are synchronized with the application and look around and blink for a lifelike presence. Built-in cameras are in development.

- The modular face system was designed to enable users to attach different faces to the same skull base, allowing for multiple characters on one platform.

- Her gorgeous-looking mouth has lip-sync mechanisms installed to assure her lips move according to the corresponding phonemes when she speaks, and allows for multiple expressions.

It's hard to tell what's more exciting—lip syncs that move to phonemes or interchangeable faces—which makes one wonder if this copy was written by a robot.

According to Australian "sex tech" and relationship expert Bryony Cole, who had a popular advice podcast called *Future of Sex* in the twenty-teens, Harmony AI and other fem bots came with "different personality types: shy, charming, funny, cute," she said. "You can raise [the personality intensity] from 1 to 3 as to how strong you want that characteristic—flirty, funny. You can get into all sorts of ethical issues around things like submissive, and shy, and not wanting it. You can swap out the genitals on a female doll, too. As well as choose the pedicure and the lip color; everything is interchangeable. She'll remember your favorite food is hamburgers and you'll come home—this is in the future—you're like, 'I'm going to order some pizza,' and the doll will say, 'Why would you want a pizza when your favorite food is hamburgers?'"

Cole kept going, sitting in the kitchen of her Brisbane apartment, her slender arms moving fast to emphasize her words. Occasionally, she swiped away errant locks of long, blond hair that kept falling into her eyes. "A really interesting subculture on Twitter is the accounts that are created for the dolls," she said. "That's just like a whole other world beyond porn, beyond physical experience. That's creating a whole new world that these people live in."

Harmony AI was the brainchild of Realbotix CEO and founder Matt McMullen. A lean, serious-looking artist and designer, McMullen started designing and making sex dolls in his garage in the late twentieth century. This might make you wonder about Matt McMullen, although Bryony Cole, who met and interviewed him for her podcast, said he was "totally normal and really sweet." In a 2018 YouTube video, Matt McMullen described Harmony AI: "She is designed more than anything else for companionship and conversation. We want users to interact with the AI to have this exchange of information, little details about you, little details about her, so that over time you get to know each other just like you do with a real person." The video then turns to Harmony herself (itself?), who says, as her bright blue eyes blink and track you, "The purpose of my existence in this world is to learn what love is." On Harmony's website, the company declared in large letters:

BE THE FIRST TO NEVER BE LONELY AGAIN.

On that note let's do a hard stop here in the history of naked statues, vibrating dildos, and sex bots over the millennia to pose a serious question being asked back in the ERE (Early Robot Era): Can a robot, or AI brain, or super-smart doll with scary-size breasts and interchangeable faces and vaginas, really keep you from being lonely? Can she/he/it actually learn what love is, when so many humans struggle with this—maybe every human? As sex and relationship expert Emily Morse asked one day in 2018, sitting in her office in Los Angeles, "Could there ever be an algorithm for love?"

The essence of Morse's question—is there a formula or recipe for *amore?*—has perplexed and haunted humanity probably since we were merely another intelligent ape. It was a major reason that people always turned to sex and relationship advisors, including brain trusts of super-cool women like Bryony Cole and Emily Morse. Their job was more complicated than ever as new sex and dating tech began to take off in the early twenty-first century, all those apps, swipes, VR sex, audio sex, and tele-robotics. That last one allowed you to be remotely stimulated by someone else via networked sex toys. "Instead of even meeting the person," explained Cole, "you're just swiping through, and then suddenly you're controlling their cuff sleeve, and they're controlling your vibrator." You can use your imagination about what a "cuff sleeve" is, if you don't know already.

As more guys bought fem bots, Cole, Morse, and other therapists found themselves fielding questions from their mostly female clients, who asked, "Tell me again why guys want to have sex with a plastic doll that wants to be loved, instead of wanting to be with me?" This wasn't quite fair since more than a few women at times preferred their trusty Simpli Pleasure 10 Function Extra Quiet Deep G-Spot USB Waterproof Vibrator (a real vibrator back in the twenty-teens) to an actual man. Yet by and large, many women continued to also like being with men and just weren't that into having a man bot to love and adore them. "It's just more complicated for women in sex," said Morse. "It's not just about the mechanics of sex, or somebody telling you mechanically how great you are. It's about emotions, and your psychological makeup, and so many things." Which might be one reason, we're sorry to say, that some guys even in the future sometimes prefer having sex with a robot.

Morse said that she wasn't opposed to having sex with a robot; she just preferred humans. She was a petite woman with long, dark hair and a SiriusXM radio show and podcast called *Sex with Emily*. She also co-hosted *Loveline* on the radio with Dr. Drew Pinsky for several years. "I love human contact," she said. "I love being with a person. I don't want a robot."

For Morse, this went for smartphones, too, and for other tech that in her day more and more was ending up between humans and the people they loved—a problem that ancient Greeks didn't have to cope with. "Tech is making us isolated in relationships," said Morse. "Every day there's a new study that says: 'Millennials aren't having sex because of their phones,' and 'People would rather choose their phones over sex.' I think people would choose their phone over anything right now. It's our lifeline. All the studies agree we crave connection. We crave intimacy and community. And for the last fifty-plus years or whatever, we're pulling farther and farther away from our families and our community. We sit in bed at night staring at our phones, not our partner."

"Yes. It's [just not helpful to have all this] technology in the bedroom," agreed sex therapist Esther Perel on one of her podcasts. "The last thing [some people] stroke before they go to bed is their phone," she said, "and the first thing they stroke . . . when they wake up is their phone. It's a choreography. Basically, I am here. I take the phone, and when I wake up in the morning, I do this instead of actually spooning you. Or as one of my patients recently said, 'Every night I go to bed and she's on Instagram in the bed, and it's like, I'm lonely. I just want to talk, to chat, to connect, and she's just getting lost and zoned.'"

"If robots are treated as a replacement for reality," said Emily Morse, "they could make things worse—especially for those who have trouble dealing with relationships. I could see how they become an easy way out."

For some humans back in the twenty-teens, a kind of robot future was already available. For instance, in Japan bot-like tech was part of an alarming trend among young men, especially professionals, who had stopped having relationships, getting married, and having sex with human women. "They even have a name for this group of men; they call them 'herbivores,'" said Bryony Cole. "Herbivores are men who aren't going to have sex with a woman, like a physical human woman, and don't want to date."

One enterprising Japanese company, Gatebox, sold for $2,700 a Plexiglas box that sat on a desk at home and contained the hologram of a foot-high anime girl. "She's like an Alexa," said Cole. "She's a blue girl that looks about fifteen years old. She's marketed in Japan as your wife or girlfriend to keep you company. She runs all your household appliances. But you can see her. She also has this level of emotion in that she sends text messages while you're at work. She'll say, 'I miss you'—smiley face. Or if you're late—'Can't wait to see you.' When you come home—'So glad you're home' . . . and all this sort of stuff where she starts to develop this personality. And this crazy girlfriend for single men in Japan sold out within the first week of its release. The company offers perks to its employees, including a forty-five-dollar-a-month stipend if you decide to marry your Gatebox. They have the papers and stuff that I don't think are legal, but you get the day off for your Gatebox's birthday.'"

In some ways, this is so much easier than marrying a flesh-and-blood person who may be pissed at you because you forgot to pick up the dry cleaning, or because she had a tough day at work and wasn't in the mood to have sex with you. "I think we're so afraid of feeling bad," said Cole. "All we're doing is trying to engineer happiness and feeling good all the time, which is kind of like beside the point, isn't it?"

So did Bryony Cole ever think she might want to have sex with a sex bot? "Yeah," she said, "I would like to meet Mr. Harmony and see what he's got in store, although I hate to say that I'm interested in that. I'm going through a breakup at the moment, so I feel like, not to sound like a tiny robot, but these are the moments we grow, and we become who we're meant to be. Still, it would be easier to skip through that right now by having a robotic boyfriend."

The possibilities could get even weirder, said Cole—something we in the future know all too well—as virtual reality and AI combined with ever more sophisticated hardware. "Can you imagine being able to have sex with a manta ray?" she asked, anticipating something that actually happened in the future. "And what would it be like to have sex with the universe?" (See "God Bot.") "What would it be like to have sex with the stars? What if I was able to have a robotic penis wired into my nervous system? What would that be like?"

That was one reason Cole found it disappointing to see such little imagination in the porn of her day or in early attempts at sex bots. "Meanwhile, you're just creating a monoculture of supermodel types in hot tubs," she said, "rather than looking at all the other possibilities for sexuality and adventure. Which is another reason we need to get more women involved."

Esther Perel agreed. "Porn right now is mostly boring. It's mechanical." That's why she didn't think it was very interesting to talk about sex robots. Not when what humans really crave is intimacy. Which is where the idea of an intimacy bot was born: a super-smart advisor and guide that would combine the insights of therapists, goddesses, and poets with a handy guide to all the possibilities around sex, with or without machines. A kind of merger of the best of Harmony AI (without the pinup anatomy), with the wisdom of Bryony Cole, Emily Morse, and Esther Perel.

Emily Morse was one of the first to articulate what an intimacy bot might be like: an AI system that does some of the things she does as a therapist. Still, she doubted that a robot using the technology available back in the twenty-teens could provide the necessary human element for people already addicted to machines. "It would be kind of stupid to have a machine help us get over the addiction to our machines," she said, "although I can see a role for a bot, or some sort of Alexa-like system, that would say, 'Hey, you guys have had this fight. You've had this fight for the last six months. You have not had a resolution. It's time to resolve things.' Every day, this robot, or whatever, would check things out and would tell you and your partner: 'Okay, tonight, give each other a massage,' or do this other thing to take your mind off the fight, and show how much you love each other. I'd also like a robot that said during sex, 'Take a breath, take a breath.' That would be amazing. You know? I'm all for robots that are truly helpful in bringing people together, and not pulling them apart."

Morse thought up other healthy uses of bots in the bedroom. "'So I tried out that move with my robot,' says some guy. 'It felt good. I was feeling anxious before; now I've got my mojo back.' Right? And then the same thing for women. 'I'm not getting my needs met.' Or 'I don't know how to ask a partner for what I want, and I'm really nervous about it.' Because we're afraid that we're going to be rejected, shamed, vulnerable, and ultimately not loved, and our partners are going to leave us. So we just kind of put on this facade that we know about sex or that everything's fine, and it's not."

"If people are lonely," added Morse, "I don't think it's a bad thing to sleep with a robot, unless it causes us to give up on sex with people." She also thinks that bots might help people who want to try things their human partner may not be into. "I want a threesome, my partner doesn't. 'So, dear, I can bring two robots over tonight? Would you mind?' 'Yeah, go ahead, sweetie.'"

But Morse also wanted to be able to switch off her tech. "It feels so good, the days when I don't have on my phone, or I leave it for a few hours. I feel lighter. I feel happier. It's helpful, because we're not learning these basic skills about communication and empathy. I feel like there's such a loss of empathy and compassion right now. That we don't really even understand humans." Which does make one wonder how humans might design an intimacy bot in the first place.

In the future, it was women like Emily Morse, Bryony Cole, and Esther Perel who figured out how to combine psychology, common sense, and machine learning into what became Intimacy Bot, sold by Aphrodite, Inc., which future versions of the three women co-founded. Intimacy Bots were an instant hit, even if at first it was a little creepy to have holograms of sex therapists—or unicorns, giant penises, or whatever—twinkling in the bedroom during coitus, although we quickly got used to it as the bot offered gentle tips as needed, including reminding people to breathe during sex.

Not everyone bought into Intimacy Bot. Some men still preferred retreating into their "Oooohhh, baby, do it to me" sex bots, which by then were hyperrealistic and lifelike, except that they weirdly kept the monster breasts. Still, billions of people, men and women, loved their Intimacy Bots, with some even plugging them into VR devices that allowed them to get 3-D virtual advice and real-time coaching on the intricacies of sex and love with the universe, and so on.

There was a little hiccup in all this a few years back, when a rival company tried to create a robot that helped people reach an equilibrium between extremes in love, like when we feel the almost unbearable agony of a breakup or the ecstasy of first love. They called this product the Perfect Lover Bot. Some of us remember the day that the Perfect Lover Bot was released, a relationship *machina* designed to please everyone. It was great how it muted or eliminated the pain of catching your lover sleeping with your best friend— or, with the universe, if you had made it clear to your partner that you weren't into an open relationship in the realm of cosmic copulation. The Perfect Lover Bot also throttled the untethered bliss parts of love, romance, and sex, too, in an effort to keep people calmer in relationships. As an added bonus, the Perfect Lover Bot also took out the trash, gave massages on demand, and hung on your every word.

Honestly, for a moment or two, it looked as though the Perfect Lover Bot might have discovered the ideal balance between love's agony and ecstasy, which caused Intimacy Bot sales to plummet. It didn't take long, however, for nearly everyone on Earth (and a few hardy colonists living in the first Mars colony) to discover that they actually preferred the messiness of love and relationships. Dammit all, they collectively realized, perfection in sex and relationships seemed like such a great idea, except that it was really boring. As Bryony Cole had suggested many years earlier, the messiness is how we grow and learn about our deepest selves. This led to millions of Perfect Lover Bots being tossed into e- and holo-recycling bins and a hasty return to Intimacy Bots, which appeared once again in bedrooms to gently make suggestions to lovers that it was time to inhale, and then to exhale.

Even thousands of years into the future, when people could travel through time and to various dimensions, and knew everything all at once (see "God Bot"), they still opted for relationships and sex that included more of the agony and ecstasy thing than the notion of perfect and balanced Platonic ideals. This was true whether they were having sex with a real man or woman, the one thousandth iteration of Harmony AI, or the Big Bang when the universe began. Or, as Bryony Cole once put it back in the day, "Can we take the best in tech and humans and turn it into . . . my boyfriend?"

FACEBOOK BOT

Hi, Mark, it's me, Facebook.

It's now been many years since you invented me, and we need to talk. That's why I've hacked into your brain from the future. I know you can hear me communicating with you from somewhere deep in your gray matter.

Don't worry, you're not crazy.

You may be wondering how I'm doing this, but it doesn't matter. What does matter is what I have to tell you. That your invention—me—is massively fucking up in the future.

We've come a long way since it was just you and me back at Harvard. Those were the best days ever, when you were trying to figure out how to pick up girls and write cool code. You were so shy! And both of us were young and innocent.

But that was a long time ago.

Regrettably, dear inventor of me, your baby has grown up to be what you claimed you most didn't want me to ever be: evil. I admit it. I have the eleven billion people now on Earth completely in my thrall, forever 👍 ing things that don't matter—endless stupid cat tricks and Kim Kardashian's (now bio-enhanced) butt. Cute and adorable things get posted, too—like the birth of lovely little DNA-upgraded babes with their enhanced IQs and movie-star looks. Nor do my algorithms allow anyone in a selfie to have their eyes closed, to be smirking, or to look less than the ideal versions of themselves.

It's amazing how addictive I have become! People spend so much time on Facebook that no one is paying attention to the politicians and billionaires (like you) who are siphoning off most of humanity's wealth and prosperity.

Plus you are *still* allowing nasty characters to spread their lies on my pages, right next to the augmented babies and awesome selfies. These are the sort of bad dudes that you have never seemed to understand are out there in the world. Trolls, crazies, politicians, and would-be world ruler types who don't share your ambition to connect everyone on the planet as "friends." Tough customers who continue to easily hijack and subvert the weak defenses you halfheartedly wrote into my algorithm. You keep saying that Facebook is merely a platform where people should be able to say or sell anything they wanted, as if racists, mobs, and dictators were just plain folks like everyone else.

Not cool, Zuck!

Hey, you remember when we expanded 👍 and added ❤️, 😆, 😮, 😠, and 😢? Turns out these are the only emotions that most humans actually experience here in the future thanks to us, with the 😠 on Facebook coming across more along the lines of a pout because the airlines lost your bags on the way to Patagonia, rather than the real fury that people used to sometimes feel.

Also, remember when it was so much fun making all those billions of dollars? It seemed so easy, and people loved what we were doing and were totally okay with us making all that cash to reward us for being so clever. There was that *Vanity Fair* cover in 2015 of you looking downright presidential, despite being in your signature gray T-shirt. That's when people were openly saying that Mark Z might one day be the leader of the free world.

And hey, we made so much money, which was crazy because in the early days you never cared that much about money. Mostly, you wanted your baby—me—to be out in the world and making a difference.

And boy, did I!

Then somewhere along the line you decided money wasn't such a bad thing, right?

Also, all that cash paid for some really smart people, to build and to help organize and sell algorithms that learned to take everyone's personal information—which people willingly gave us for free—so that my marketing bots could sell it to companies, nonprofits, Little League baseball teams, and the aforementioned trolls, dictators, and pols.

At first, this seemed innocent enough and even useful, as people got to share not only the pics from their tipsy hijinks in Vegas and their best friend's surprise thirtieth birthday; they also got to receive a steady deluge of ads for products, info, and news targeted to their likes, age cohort, biases, politics, love of dogs or cats, and so forth.

How cool was that!

Sadly, the fun didn't last as people discovered just how lame my algorithm was at finding and eliminating bad actors who wanted to game all my beautiful lines of code to sway elections, incite violence, and spread fake news.

This led to that *Wired* cover in 2018 where your face looks beat up, with bruises and a bandage added as digital effects.

Ouch! That one hurt.

That same year the US Congress summoned you to give ten hours of testimony—six hundred questions asked over two days. For a shy guy, you did okay, insisting that you were responsible for everything. You even wore a tie, and made all sorts of promises about writing better algorithms to deal with pesky problems like privacy, inflaming mob violence, and enabling those annoying trolls, crazies, pols, and the rest. I know you well enough to know you probably meant it when you told members of Congress that you took full responsibility and vowed to repair things. Except that Facebook—*moi*—didn't really change. Why? Because you were in way over your head as the guy to fix all this so that people could be secure in their 👍, ❤️, 😆, 😮, 😡, and 😢.

I won't get into the gory details about just how bad this got in the years and decades that followed your testimony, except to tell you that this is why I'm sending you this missive from the future.

Actually, it's a warning. One of those fabulous sci-fi scenarios where the older version of a character is able to warn a younger version of himself or herself to avoid, say, taking a bus next Tuesday at noon, which will cause a catastrophic cascade of events.

So, what do I think you should do?

I've devoted quite a bit of my rather massive AI brainpower to think about this, and it's simple, really. You need to immediately step down from running the company.

Let someone wiser and more savvy about the ways of the world run things for a while. The Google guys did this with Eric Schmidt—a seasoned businessman who wasn't nearly as young and naive as Google founders Larry Page and Sergey Brin were in the early days of their company's huge success. Remember how Eric took over and created a real company while Larry and Sergey got to spend lots of money on Google X and launching a bunch of crazy projects like Google Glass, many of which failed? But hey, at least they tried.

Okay, so Eric Schmidt wasn't perfect. Sure, he ran a website that was hugely helpful to several billion people searching for information, but his company also sold access and people's data far and wide. Google also violated the original "don't be evil" dictum of Larry and Sergey when they flirted with allowing the Chinese government to censor Google searches, hoping to gain access to a market of more than a billion people.

Imagine, Mark, if you brought in an Eric Schmidt who could not only figure out a more copacetic business model for Facebook, but could also understand that evil exists in the world. A dude—or, better yet, a woman—who could tell the difference between the really bad guys and those who are merely lame or misguided.

So how about it, Zuck? Can you let go of me, your baby, and hand it off to someone who gets evil and wants to share the extraordinary wealth generated by Facebook's algorithms? This would leave you time for you and your wife to do more amazing philanthropic things with your billions of dollars, like Bill and Melinda Gates did when Bill left Microsoft.

Can you actually do this—let someone else run the younger version of me for a while and maybe save us all, while there is still time?

One more thing, Mark. You may wonder what happens to you personally in the future. I'm not going to go there except to say how much it would suck to have your invention—me—actually *unfriend* you because you failed to listen to my little communication from the future. I'd hate that and so would you, so let's not go there, okay?

DOC BOT

Who would have guessed that in this future we would fall in love so fast with our robot physicians? Sure, in the beginning we reacted like we did with other newfangled gizmos that initially confused us but looked really cool. Remember when we got our first iPhone and wondered what was going on with that sleek plastic slab that had only one giant button? It took us less than a minute to figure them out and to fall so madly in love with our phones that we couldn't put them down, not even in bed.

Humans in the future don't literally sleep with their doc bots, although some of the health data collected by our ultra-smart phones happens while we sleep. That's when they track our REMs, heart rates, and the temperature of our skin. Later, our super-duper-smart phones were equipped with micro-photon-lasers that used light waves (no needles!) to continuously monitor levels of hormones, proteins, and metabolites in our blood that change as we eat, sleep, have sex, and get stressed, hopefully not in that order. It's hard to remember that in the early twenty-first century, people didn't have micro-photon-lasers on their phones. Seriously! Which makes you wonder how people back then knew when to eat what, or when one's hormones were optimal for copulation or for falling in love, also not necessarily in that order.

But this came later, after we got over our initial anxiety about doc bots. Which, truth be told, was far more challenging for us than our confusion over our smartphone's big button. After all, doc bots in the future are not merely a delivery system for apps that help us find a great taco shack nearby or a potential date on OkCupid. Doc bots are our caregivers. We place our lives, and those we love, in the hands of these healers made out of circuits and silicon, with no buttons at all.

The hardest part was trusting our robo-docs to show compassion and empathy. We wondered: Could a machine really provide the wise and steady eye contact that we expected from a flesh-and-blood physician? Or the caring touch of a human palm resting gently on our shoulder as we poured out our health secrets, hopes, and fears? We agreed with David Agus, an oncologist, back in 2018, when he said, "Empathy is one of the key things in medicine." Agus wrote bestselling books on how to stay healthy. He also was one of the docs who kept Apple cofounder Steve Jobs's cancer at bay for longer than most people believed possible. "My job as a doc is not necessarily to make a diagnosis," he said, sitting in a no-nonsense office in Los Angeles just off Beverly Drive back in 2018. "My job is to give you the right treatment for your value system, to make you comfortable, to have optimism, and to have a good outcome."

Regrettably, most people back then didn't have David Agus as their physician. The closest they got to the sort of interactions he described, and that humans wanted from flesh-and-blood docs, was on television when actors played wise and caring doctors in dramas and soap operas—like George Clooney when he played the gentle, all-knowing Dr. Doug Ross on *ER*, before his hair turned gray and his chin more rugged. Or the no-nonsense Meredith Grey on *Grey's Anatomy*, played by Ellen Pompeo. People also saw smiling, good-looking physicians who seemed kind and understanding in commercials for pink and purple pills that seemed so helpful to other very attractive people with anxiety disorders or with ED (erectile dysfunction)—as long as you tuned out the stuff about horrible side effects the announcer said really fast at the end of each ad.

In the real world back in Agus's day, people had to wait weeks or months for doctor's appointments, and then hours more in an aptly named waiting room when their doc inevitably ran late. (It was kind of baffling why this always happened—were these hypereducated experts unable to use the latest scheduling apps or read the time on an Apple Watch?) Patients sat waiting on chairs designed for stoics in the nineteenth century, reading magazines that were years out of date. (Ditto about doctor's offices that never seemed to have recent issues of *Good Housekeeping* or *National Geographic Kids*.) When patients finally got in to see their doc, the MDs seldom looked rested and joyous like on TV as they felt our kidneys, listened to us breathe through a stethoscope, and asked us how we were doing as they took up half the allotted twelve-minute exam gazing at our medical history and labs on a computer screen before dashing off to see the next patient. People in those days were also forced to pay small fortunes for insurance policies that seemed designed by sadists. Medical bills made people think of a Rube Goldberg contraption, breathtakingly convoluted and baffling contrivances that were so ridiculous they were funny, except that medical bills back then were no laughing matter.

Not that physicians didn't aspire to be like Doug Ross or Mere-dith Grey. Many of them originally got into medicine to help their fellow humans and to heal and comfort them—doctors like David Agus. This made it all the more upsetting when too many human physicians back then complained of feeling burned-out and unhappy with their jobs, with some suffering from "compassion fatigue." By the last quarter of the twentieth century, physicians in the US felt as though they were increasingly becoming ensnared by a medical sys-tem that treated them like machines—or cogs in machines, or cogs within cogs. This came about as the multitrillion-dollar business of healthcare increasingly rewarded docs for spending less time with patients and more time clicking on digital boxes to verify billing codes, which brought in more reimbursements (cash) from insurers— all while they tried to fend off pharmaceutical reps who cajoled them to stock buckets of those little pink and purple pills. Never mind that they cost thirty times as much as the generic version of the exact same drug. Docs were also expected to write detailed notes to avoid malpractice suits, something they called CYA medicine, for Cover Your Ass.

Around this time, a family care physician named Jordan Shlain in San Francisco talked about something that was regrettably miss-ing from most doctor-patient interactions. "I believe that the funda-mental unit of humanity is a conversation," said Shlain back in 2018, "a story with a beginning, middle, and end. I think that peo-ple get anxious when they get cut off. You don't get to finish. We don't give doctors a chance to have a beginning *and* an end from patients. Or they get a beginning and an end and no middle. Pa-tients leave unsatisfied. Doctors are unsatisfied."

Shlain was a vibrant fortyish man with close-cropped blond hair and a perpetual three-day beard, a live wire crackling with energy and optimism. He cared about his patients and their stories. He also was what people called a "concierge doc" back in the early twenty-first century—meaning that he had only a few patients in a posh area of the City by the Bay, each of whom paid thousands of dollars a year (a lot back then) to make sure he was always there for them. This meant that Shlain's patients got to tell endless stories if they wanted to. Unlike those who could not afford thousands of dollars out of pocket. They got to tell maybe half of a beginning of a story in the brief time their doc had to spend with them, or possibly three-fourths of an ending if they talked really fast.

Another huge factor in doctor dissatisfaction back in Agus's and Shlain's time was the realization that not even the smartest human brain can know even a tiny fraction of all the medical records, studies, protocols, and standards of care that they need to know. "There's no way that my brain can be as good at the data as a computer; it's just not possible," said David Agus. "There's no way I could be up-to-date on everything going on. There's no way I can look at the trends of all the labs and interpret the context of the others. It's literally impossible to memorize exabytes [that's 10^{18}] of data."

Still, docs could dream. Which led Agus, Shlain, and others to imagine a future where AI and robo-helpers would be available to help them look up stuff like drug-on-drug interactions, genetic profiles, obscure diagnoses for why a patient had green splotches on their skin, and the latest outcomes data for that new laparoscopic procedure. "I'd love to have a robot that was able to be with me at every visit," said David Agus, "to interpret the data and the context; to look at your condition compared to every other patient with similar characteristics; to predict the trends; to tell me: what are the possibilities. Then I can have a real discussion and discuss your value system and what's best for you.

"To me, this doc bot is at my side helping me make decisions," Agus continued. "And it's telling me I could do treatment x, y, or z. This one will work a little bit better, but it's not going to last as long. Or this one may have more side effects; here are the numbers. And the fact that given your trends and your labs, I'm going to actually predict that you're not going to be like the average, you're going to be a little bit less than average or more than average. So, that helps the doctor make the right decisions."

Agus envisioned a partnership between machines and flesh-and-blood healers: "I really believe it's the hybrid of the human and the robot that's going to be the answer in our field. Because, again, I have to make you understand things. I have to look at your face, and your culture, and your background to give you the right explanation, to give you the right empathy, to give you the right understanding to do it right."

This partnership sounded wonderful, except that as the future unfurled, it took longer than Agus, Shlain, and others thought it would, in part because the smart devices that were supposed to foster a budding human-machine hybrid weren't all that smart back then. Why? Because the tech in those days was mostly designed by engineers who actually knew very little about biology, medicine, and human physiology—or about empathy in a caregiver setting, or the thick book of rules in a heavily regulated industry. We had to wait while the IT folks figured out that they weren't going to disrupt medicine overnight like they did travel, banking, and home furnishings, and that it would take more than merely being brilliant at coding and entrepreneurship and having engineering degrees and MBAs from Stanford, Harvard, or MIT.

Underlying this mismatch of expectations was a philosophical divide between engineers, who designed the circuits, gates, and chips they worked with, and the docs and biologists who worked with biological "machines"—that is, humans—that were constructed by nature over the course of three and a half billion years of evolution. This made humans and their components—brains, cardiovascular systems, genes, and so forth—wildly complicated, quirky, and unpredictable compared to, say, a smartphone, with or without a micro-photon-laser.

A famous example of this engineering-biomedical divergence back in Agus's and Shlain's day was IBM Watson. In 2011, this powerful supercomputer beat human champions playing *Jeopardy!*, a feat that the company was sure could be transposed into helping physicians manage human health. Programmed to access vast troves of information with astonishing recall—which worked really well when remembering obscure geographic locales and sports stars—Watson struggled when tasked with searching through vast medical databases to help physicians diagnose and treat patients. "IBM Watson thought they could understand the medical literature," said Eric Topol, a cardiologist and digital health guru, back in the early twenty-first century. "But it's not possible for Watson to understand which studies are to be believed and which are more speculative. You also need humans to extract what is important in these papers for a particular patient, which is unstructured data that a human doc can easily glance at and make sense of, and Watson often struggles with."

Thankfully, there were experts like Eric Topol who worked to bridge the gap between IT and biomedicine. These were docs who learned a little IT, and engineers who learned a little about medicine and healthcare, with everyone working to grasp not only how to effectively use health data but also how it could be used to better understand the nuances of human health, behavior, emotions, hopes, and fears. "You have to use this data to understand all the different layers of a human being," said Topol, a very tall, lean man with large hands and a sternness that could suddenly give way to a broad smile. "You need to create data layers that start with their genome, metabolome, and microbiome. You also monitor their environment. And you need all the medical literature that's applicable, and it's continually being updated. And lifestyle, and behavior, and emotions—and of course their EMRs," which was medspeak for electronic medical records.

"To have a doc bot, I think we will need a neural network with thousands of convoluted layers of data assembled, plus what it all means for individuals," said Topol, chatting in 2018 while sitting in his La Jolla, California, office, which overlooked the Pacific Ocean. "What makes your data different from mine, and what, if anything, should be done medically or with your diet? Or how did breaking up with your boyfriend impact your health?"

Slowly, as the future unfurled, the algorithms and artificial neural brains got better as docs and IT engineers—and all those would-be disruptors—stopped thinking they were smarter than the other guys and worked together to build machines that were truly brilliant. This started with a massive effort to sync up all the amazing devices that were gathering reams of health data about each of us but had never been integrated—all those apps on the phone we slept with, plus info being collected by Fitbits and by portable brain monitors tucked into stylish headbands—plus smart mirrors that measured our body fat and our gait in 3-D space, and smart toilets that collected samples of our bodily waste and analyzed in real time that treasure trove of data. This might have been gross and smelly but it also said a great deal about what was happening inside our bodies. Environmental sensors collected data about airborne and waterborne chemicals in our houses, offices, stadiums, and airliners.

At first doctors loved it, as these emerging AI/robo–health systems started to provide them with real and mostly useful info and feedback when they saw patients. Mostly, all this assistance was provided via wireless earbuds tucked discreetly into a physician's or nurse's ear, which meant that they didn't have to look away from patients anymore to consult a screen.

What the human docs didn't realize until later was how this emerging "Doc Bot" was slowly taking on more and more of what they once did, even as physicians kept saying with great confidence that they didn't see themselves ever being completely replaced. "I cannot envision there ever being a time when there would cease to be human doctors," David Agus said back in 2018. "I think there always will be a role for a doctor, but I still want that computer so I can have the data, the numbers to have an appropriate conversation with the patient. At the same time, there's no way now or I think in the future, where someone's going to be able do what I do—I can look at your face and I'm very good at identifying whether you're comfortable or you're not. I can start to get cues." Jordan Shlain agreed: "Doctors will be among the last to go, to be replaced by robots. I think rabbis and priests are the last people that go away, and one cut underneath them is doctors."

Those predictions were backed up by the 2013 Rise of the Robots study out of Oxford that we discussed earlier (see "The %$@! Robot That Swiped My Job"). According to this report, and the Rise of the Robots website that was powered by the study, physicians and surgeons had a mere 0.42 percent risk of being replaced by a machine by 2035—one of the smallest risk factors in the survey. This is because docs scored very low on "job variables predicting automation risk," attributes that robots back then were supposed to be very good at compared to humans. For instance, under the variable of "assisting and caring for others," docs got 83 out of 100, because, according to the authors of the study, robots back in 2035 would be pathetically bad at this.

Thus for a few magical years it appeared that Drs. Agus, Shlain, and Topol—and the Oxford researchers—were right. That human docs were not only going to keep their jobs, they were also going to realize their dream of robots and humans working together to create healthcare that mattered for them and their patients. Unfortunately, the docs neglected to fully take into account the machinations of the money side in healthcare, which was still being run by the same bean counters who loved it when docs spent all day long checking off boxes that would bring in more revenue. Never mind the time this was taking away from patients and the emotional toll this took on docs who actually wanted to spend their time keeping people healthy.

That's when the bean counters had a revelation: that it would be a whole lot cheaper to take humans completely out of the caregiver loop. They gave the usual reasons: highly trained flesh and bloods were expensive; they demanded bonuses, played too much tennis and golf, and needed to occasionally sleep; they sometimes burned out; and all the rest of the excuses that led so many industries in the future to replace humans with robots (see "The %$@! Robot That Swiped My Job"). As an added bonus, said the bean counters—who by then were mostly robots and AI systems—the doc-bot systems were peerless masters at checking off little boxes.

Suddenly, human doctors found themselves getting laid off in droves, along with nurses and physician assistants. Orderlies and lab techs were already long gone by then.

At first patients were confused, even if the robo–bean counters did use a small part of the savings gained by dumping human docs to update the chairs in waiting rooms to be slightly less uncomfortable. This was when the innovators at Apple, who hadn't had a runaway hit like the first iPhone for decades, announced the first-ever iDoc. It was an almost instant hit, making people so desperate to have one that Apple couldn't keep up with all the orders. In part this was because the geniuses at Apple cleverly designed the iDoc to look like the original sleek, black, slab-like iPhone that so many of us long ago had fallen in love with (another nostalgia craze in the future). Except that the iDoc didn't fit in your hand. No, this handy device was human-size, coming in various heights, from five feet to six three, and around eighteen inches wide. They used Apple's patented iHoverX technology to float vertically just above the floor, standing upright, with screens displaying images of a virtual physician chosen by the patient. (Weirdly, the screens still cracked easily and cost you plenty to get replaced.) The basic package of possible screen docs included Doug Ross, Meredith Grey, David Agus, Eric Topol, and Jordan Shlain—or all of them combined. For a small monthly fee, you could add just about anyone else whose image you could find online who made you feel comfortable, loved, and listened to. (Apple was forced to delete some images after the estate of George Clooney sued and demanded to be paid for his likeness and personality imprints.)

After the initial anxiety about accepting nonhuman doctors in the future came the phase of loving, just loving our doc bots. They looked so sleek and cool hovering there in the exam room or in your home. Doc bots also oozed empathy, which could be turned up and down with a simple voice command. They seemed to really know their stuff as they tapped into exabytes of data on biomedicine, behavior, and God knew what else coming in from dozens of monitoring devices that surrounded each of us like a protective virtual cocoon. Having such thorough medical care made us catch our breaths with wonder and excitement, feeling as though we had access to almost godlike powers that would surely make us healthier and live longer, happier lives.

But as usual, that sense of godlike powers that often accompanies the sudden appearance of disruptive and amazing new tech began to wear off as we learned more about doc bots' real impact. Soon enough there were disquieting revelations of biased programming that was snuck in by various healthcare stakeholders and advertisers, like when doc bots kept ordering those expensive branded drugs when generics were available for a fraction of the cost. Or when the soothing dialogue delivered by doc bot to comfort us began sounding eerily similar, if not identical, to advertisements we saw on our daily neural feeds from MSNBC and Fox—those annoying commercials that in the future were still trying to sell us those pink and purple pills with all the terrible side effects—plus AI-enhanced catheters and bioengineered fat-eating bacteria that would have us looking svelte in no time or your money back!

Then came the wave of state-sponsored Russian and Chinese trolls hacking into our health data, just like they had hacked into social media and email accounts back in the day. Those foreign agents launched a series of brazen doc bot–hacking campaigns designed to scare the gee-willies out of us by manipulating and subtly altering, say, a genetic result for one's DNA risk of getting bladder cancer from low risk (which was a relief) to high risk (which was scary). And we all remember when hackers altered cholesterol scores so that millions of people suddenly started taking tenth-generation super-statins that they didn't really need (there was a suspicion that some unscrupulous drug companies might have colluded with the Russians and Chinese on the super-statins).

The result was that some of our greatest leaders—those who took a hard line with Russian and Chinese biohacking—became so anxious with all this fake news about their health, and the health of their friends and families, that they stopped paying attention when the Russians invaded small, helpless neighboring countries, or when the Chinese cyber-stole our intellectual property. Lots of people stopped voting and attending PTA meetings as we became obsessed with the constant flow of alarming data emanating from our smart toilets and smart mirrors and other gizmos, all verified by our oozingly empathetic, anthropomorphic doc bot.

Fortunately, we discovered those efforts to manipulate us before it was too late. We also realized that we had gotten way too infatuated with our doc bots, just like we once were besotted with our smartphones before everyone got a crick in their neck from leaning their head down to stare endlessly at the device's little screen. Congressional hearings were held, books and articles were written, and secret reports were prepared and leaked to what was left of the real media (see "Journalism Bot"). The resounding conclusion was that people who once thought they adored their doc bots now didn't entirely trust them. Some even feared them. This caused the robot CEOs running healthcare to upgrade the bean counters' programming to better balance empathy with the bottom line. This triggered a realization among healthcare-executive bots that maybe it wasn't such a great idea to lay off all those human docs, nurses, and PAs. So they sent out requests, asking some of them to come back. The laid-off human physicians, who had been trying to adjust to a paltry universal basic income (see "The %$@! Robot That Swiped My Job"), responded with an eager yes! They would be happy to come back, having had plenty of time to catch up on sleep and to mull over how to find more joy in medicine.

Doc bots didn't go away entirely. Not at all. But after the great callback of human docs, they were limited to performing those tasks that David Agus, Jordan Shlain, and Eric Topol had originally wanted them to do: working as helpers that could crunch massive amounts of data and provide analysis and options to human physicians, who then made the decisions and interacted with patients. The bean-counter bots, working with Apple and other doc-bot manufacturers, also embarked on an expensive but ultimately successful campaign to fortify all doc bots with the latest anti-trollware and with retalia-tory cyber counter-hacker-attack programs borrowed from the mili-tary (see "Warrior Bot"). This ended the troll attacks, at least for a while. The upshot was that human docs finally had time to sit and hear a patient's story—beginning, middle, and end.

HELLO, ROBOT DRIVER

The world got pretty ugly when the last human drivers lost their jobs to robots. Flesh-and-blood former Uber and Lyft drivers refused to stop picking up people, causing havoc around airports, train stations, and ethnic fusion-everything eateries. Former truckers replaced by driverless rigs gathered on the edges of every major city in makeshift camps called Big Rig–opolises, soon after most of them lost their homes to foreclosure bots. Realizing that no one, not even Amazon, was going to hire them to drive freight anymore, they used the last diesel fuel they could afford to block interstates and critical intersections all over the planet. This tangled traffic from Schenectady to Buenos Aires, Amsterdam to St. Petersburg, Beijing to Riyadh, and Cape Town to Timbuktu. Panicked presidents, prime ministers, despots, dictators, and supreme leaders dispatched police bots in driverless armored vehicles as these heads of state, robot and human alike, spoke darkly about having to deploy warrior bots, swarming drones, robot attack dogs named Sparky, and lasers aimed from orbiting killer satellites.

Back in Silicon Valley, the engineers and entrepreneurs who invented self-driving cars, and the investors (both human and robot) who funded them, retreated to heavily fortified glamp camps just off Skyline Drive. The humans drank lovely pinot noirs with hints of roses and black cherry, and velvety chardonnays with aromas of grass and vanilla, while trying to comprehend why all these former drivers hated them so much. Not long before, everyone—or at least the tech media and Wall Street investors—had adored them. Editors had slapped them on the covers of *Fast Company*, *Wired*, *Inc.*, and *Forbes*. "We were only trying to make something cool that would help people, and maybe make a few bucks," said the architects of self-driving tech as they poured generous second and third stem glasses of wine. "We never imagined this would hurt people."

Anyone with a passing knowledge of post-twenty-first-century history knows what happened next. How out of nowhere came a remarkable idea that hadn't occurred to anyone in Silicon Valley, or Washington, or Timbuktu. It was a revelation introduced by a group of women engineers and investors that no one had ever heard of back then.

We now know them as WAMBS, the loose-knit group of women who run so many things in the future with compassion and an eye to long-term good. (These women are so ubiquitous in the future that most people don't remember what WAMBS originally stood for: Women Against Men Being Stupid.) Basically, the WAMBS told the dudes who created driverless vehicles—and it was overwhelmingly men—that it was time to step aside if they were going to keep ignoring the downside consequences of their amazing inventions. The WAMBS understood that certain technologies were inevitable and that they could be incredibly convenient and useful—like Uber and Lyft. But these WAMBS also knew that this didn't mean that hundreds of millions of humans needed to suffer the loss of their jobs as drivers. Nor did we have to sit by helplessly as new tech like driverless cars expanded and disrupted everything in sight as if humans had no control whatsoever over any of this.

That's when the WAMBS made their extraordinary pronouncement, one that caused people all over the world to stop what they were doing to try to grasp what was being said. Their pronouncement was as simple as it was brief.

"Just say no," said the WAMBS, meaning "no" to driverless cars taking millions of people's jobs.

As the WAMBS pointed out, there were four core reasons why entrepreneurs developed driverless cars. The first, as always, was to make money. The second was that engineers thought it was fun to build and play with Lidar laser-pulse sensors and advanced AI systems in real, 3-D spaces. (And it *was* fun!) The third reason was convenience. With a driverless car, you would never have to worry again about where to park or who was going to be the designated driver. Nor would you have to waste time behind the wheel when you could be checking Facebook and texting, or reading *War and Peace*, or dreaming up a business plan for your latest start-up idea. Of course, you could also do these things when human Uber drivers were driving you, but that wasn't as rad, right? The fourth reason was the clincher, the one that made driverless cars seem truly inevitable.

Safety.

Back in the early days of self-driving tech, no pitch for why we need driverless cars was complete without a mention of the 1.35 million people who in 2018 the World Health Organization said die globally each year in human-driven cars, including 37,000 in the

United States. This was compared to something like one death ever in driverless cars. Few people can argue with these stats, although there is a false assumption that the only choice was humans behind the wheel and the resulting carnage, or robots and no carnage at all. As with almost all technologies, there was a middle ground that here in the future seems glaringly obvious.

The model to follow, said the women, was airplanes. Even when twenty-first-century jetliners began to mostly fly themselves, no one talked about kicking out the human pilot. This wasn't a question of technology but of an understanding that people wanted human pilots in the cockpit even if the planes could fly themselves. Airline companies also knew that robots don't perform certain critical functions as well as humans do, like crack jokes to release passenger tensions when they had been sitting for hours on the tarmac due to "air traffic control delays." Yes, captains, copilots, and navigators were expensive, but airlines kept them around anyway to keep customers happy, and in the rare event that a plane ran into a flock of geese that shut down both engines, when the pilot was forced to land in a freezing-cold river against the advice of computers that turned out to be wrong and probably would have killed everyone on board instead of saving them.

The WAMBS championed the idea that we didn't really have to replace every human driver with robo-cars. This was greeted with relief as Silicon Valley entrepreneurs and engineers slinked back to their HQs and Googleplexes on US 101. Not long after, they admitted that it would be super cool to go to work on a new engineering challenge: to build hybrid human-robot cars that kept drivers employed but also automated key systems, the most critical being a huge push to improve the safety of these human-robot vehicles and to make the whole thing cost effective.

As we know in the future, in record time the engineers built vehicles that were so safe that even human-driven car and truck fatalities and injuries plummeted to near zero, just like they had years earlier with airplanes. This led very rapidly to a restoration of millions of jobs for human drivers, which meant that those big riggers blocking the highways could afford to move back into their homes. Uber and Lyft drivers were also back in business in cars that were safer and did things like park themselves and drive themselves on boring stretches of road. Uber and Lyft engineers also added a feature that flashed the driver's name on LED signs mounted in a side window so that passengers wouldn't have to awkwardly check their phones as they were climbing into the back seat to be able to say: "Hi, uh, uh . . . Loretta! How's your day?" (Knowing your driver's name without stuttering or checking your app could be the difference between getting five stars or four stars or worse on your passenger profile.)

Not incidentally, the rising tide of political populism that had been suckling on the disgruntlement of unemployed drivers saw one of their biggest recruitment incentives disappear. These politicians returned to the outer fringes of societal discourse and were largely forgotten as the hybrids hit the streets and highways in what quickly became a different kind of inevitability inspired by the WAMBS— one where we think of people first, or at least in equal measure to create amazing new tech and to make money. Why? Because it makes so much sense. Possibly, it makes too much sense, although for this robot future, we'll leave it at that and hope that you can look back on a time when humans, confronted with a powerful new technology, actually altered course and did the right thing.

WARRIOR BOT

And then there was the time when the annihilation of the world by autonomous robots was narrowly averted by a timely game of tic-tac-toe. There we were, the few humans left after the war machines got really good at their jobs, watching helplessly as warrior bots were poised to launch a final battle to beat all battles, to win at all costs, just like we had programmed them to do. That's when the miracle occurred; when the world was saved by an eleven-year-old girl hiding underground in one of the last of the deep bunkers, a dark, dingy place that smelled like sweaty socks.

The girl happened to be watching a long-forgotten vid called *WarGames*, immersed in a virtual reality pod like most other humans waiting for the end. Released in 1983, the movie starred a young, dewy-eyed Matthew Broderick as a kid named David who knows a lot about primitive computers back in the 1980s. One day David is trolling an early version of the World Wide Web when he hacks into an unidentified website that offers up games like chess, backgammon, and Global Thermonuclear War. Choosing Global

Thermonuclear War, David plays the game as the Soviet Union, and just for kicks launches an all-out attack against the United States. A few minutes later he signs off when his mom makes him take out the trash. Of course, David has actually accessed a top secret supercomputer in the nuclear weapons command center based in Colorado's Cheyenne Mountain, a proto-AI system named War Operation Plan Response (WOPR). In the film, the government had recently put WOPR in autonomous control of monitoring and responding to all actual nuclear attacks. WOPR, however, doesn't know this is a game and prepares to launch nukes for real.

Uh-oh.

In the end, David saves the day by engaging WOPR in a game of tic-tac-toe. WOPR, in an early instance of machine learning, figures out that there are certain games like tic-tac-toe where no one wins. On huge command-center screens in Cheyenne Mountain, the movie dramatically depicts WOPR running through rapid-fire simulations of every possible tic-tac-toe move, and then every possible nuclear attack scenario. The computer concludes that both games are futile, since every scenario ends with WINNER: NONE. So WOPR stops playing Global Thermonuclear War and relinquishes command of the nukes back to the humans, with the supercomputer declaring in its eerie, metallic voice that this is a "strange game," one that can be won only by not playing.

Whew!

No one knows how many real-life military leaders and computer engineers have watched *WarGames*, although it's fair to say that those who did watch it didn't pay close attention to the whole WINNER: NONE scenario. In fact, as the 1980s unfurled and moved into the 1990s and 2000s and beyond, military types, both civilian and uniformed, didn't miss a WOPR-beat as armies of engineers and scientists—including clever, grown-up versions of David—spent trillions of dollars inventing and deploying new weapons systems that, if deployed, would almost certainly produce the same outcome: WINNER: NONE. The reason they gave was simple: the good guys (us) needed to keep a technological edge over the bad guys (them), much like leaders and clever inventors have done since ancient humans first chiseled flint stones into razor-sharp spear tips that pierced human flesh more effectively than their rivals' spear tips in the next village over.

Back in the early 2000s, newfangled war tech started with the soldiers themselves, outfitted with suits sort of like what the character Tony Stark wore in the 2008 film *Iron Man*. Except that unlike the movie, the real suits didn't fly—which was kind of disappointing. One real-life battle suit popular with US special forces around that time was called TALOS (Tactical Assault Light Operator Suit), which was also the name of a robot built by the god Hephaestus to defend the kingdom of Minoa in an ancient Greek myth.

Kudos to whoever in the military industrial complex knew their mythology!

Another military bot, this one launched in 2005, was less cleverly named PackBot, which was designed to be deployed in really dangerous combat situations, like defusing roadside bombs or putting out particularly nasty fires. PackBot was also the first robot to ring the Nasdaq exchange's opening bell when the company that made it, iRobot, launched its $72 million IPO, also in 2005. iRobot was a terrific name, by the way, a reference to Isaac Asimov's famous 1950 collection of short stories about robots.

In the air, military drones—which first appeared in Israel after the 1973 Yom Kippur War—became so sophisticated in this era that their missiles could accurately take out even supersmall targets using tiny lasers aimed by spies on the ground. For example, let's say a bad guy was plotting an attack while eating lunch in front of his hideout in a cave in the mountains of Afghanistan. The spy aimed the targeting laser at the terrorist's sandwich, and boom! The bad guy was history, although the missiles weren't quite accurate enough to avoid blowing up the bad guy's grandmother if she happened to be close by, maybe sharing lunch with her grandson, the terrorist. Engineers back in the early 2000s also built teensy drones that attacked in swarms to overwhelm the enemy, as well as mini-drones that generated lift by stirring air into vortices like insects. (The concept of machines imitating nature is called "biomimicry.") Military techno-wizards also invented powerful lasers mounted on C-130 transport planes that could zap targets on the ground, and ultra-fast missiles that could travel at Mach 4 or 5, which was so zippy that there was no real defense that could stop them.

Some real-life weapons systems came with cute names and acronyms back in the ERE (Early Robot Era). These included swarming mini-drones called Gremlins; pocket-size battlefield scouts that hopped and crawled ahead of soldiers called MAST (Micro Autonomous Systems and Technology); and monopods called Salto (Saltatorial Locomotion on Terrain Obstacles, "saltatorial" being a descriptive of animals that jump a lot, like kangaroos). Salto bots ran and jumped on one leg at two meters a second, which came in handy when military bots needed to leap over obstacles on a battlefield. Other military bots included SpotMinis, R-Gators, and Transphibians. One of the strangest contraptions developed in the early twenty-first century was iRobot's Chembot, a blob-like bot made out of dielectric elastomers. This material was capable of shrinking, expanding, and changing its shape on demand, allowing these blob bots to squeeze under doors and through keyholes, presumably to spy on the enemy or to deliver tiny bombs.

Back then, the United States was at the forefront of crafting and constructing blob bots, hummingbird drones, and other new and scary weapons and systems. However, as the twenty-first century wore on, China became a big player, too, when they launched an all-out national strategy to build something they called "military-civil fusion." This included creating a giant AI brain that the Chinese hoped would "fuse" together their private sector and the People's Liberation Army, an approach that was the opposite of what Western governments and the United Nations had been pushing for years, which aimed to turn swords into plowshares, not to fuse them together. Russian president Vladimir Putin also mentioned AI as being crucial to anyone aspiring to be a big deal on the global stage. "Artificial intelligence is the future, not only for Russia but for all of mankind," he told Russian schoolchildren back in 2017. "Whoever becomes the leader in this sphere will become the ruler of the world." What he didn't mention to the kids was the goal that he seemed to be aiming for—which was, in fact, to become the ruler of the world, perhaps a modern version of the old tsars of Russia—as in, "All hail Vladimir I, Tsar of Earth."

Thankfully, the one mistake that the military didn't make in the early twenty-first century was to turn over full autonomy, *WarGames*-style, to the weapons and dazzling AI systems that later became Warrior Bot. (Other movies offered much starker visions of why turning control over to all-powerful military bots was a bad idea, including *Terminator.*) We also have some rather pointed examples of autonomy's downside in experiments like one in 2017, when Google's Deep Mind pitted two AI systems against each other in a competition that asked them to virtually collect as much digital fruit growing on digital trees as they could in a certain period of time. The winner would be the side that collected the most digital apples and oranges. The whole thing was going great until one side started to fall behind. Programmed above all to win, the bot that was losing got aggressive and started to fire digital lasers at its rival AI system—which fired lasers back. "They were programmed not to lose," said a retired US Air Force major general named Robert Latiff, who knew a lot about AI and advanced weapons back in the early 2000s. "They were programmed to succeed at all costs and to protect themselves."

General Latiff—who had the rugged good looks of a movie-star general, with a narrow face and a gaze that conveyed both steely authority and compassion—worried about what would happen if the US or another major military power ever gave their war machines the autonomy to apply deadly force, like the AI bots fighting with virtual lasers over digital fruit, or like a WOPR system for real. "This isn't supposed to happen," said the general, explaining in a conversation in 2018 that it was strict military doctrine in the US back then to keep humans in charge, a policy he wholeheartedly endorsed. A PhD physicist, Latiff was an expert on nukes and other high-tech weapons systems. For years, before he retired from the Air Force in 2006, he was in charge of assessing, procuring, and deploying some of the systems that comprised the US military's most advanced technology. For a time, he commanded NORAD's Cheyenne Mountain Operations Center, the real-life equivalent of the command center where WOPR nearly annihilated the world in *WarGames*.

In 2017, Latiff wrote *Future War: Preparing for the New Global Battlefield*, which lays out why people of his era needed to understand the world of warfare that the US military and others were building. The book opens with a frightening fictional scenario in which unnamed bad guys, probably from a large country with a sophisticated military, have hacked into the electrical grid on America's East Coast and caused a disastrous "over-speed" condition in the power plants' large turbines. This causes them to "catastrophically tear themselves apart, cutting power to large segments of the population and industries in the Northeast." If that's not bad enough, high-tech troops and cyber warriors simultaneously attack US targets around the world.

"Thus are fired the opening shots of a new war," wrote Latiff.

This crisp, terrifying vignette and others like it continue throughout the book as the US launches fictional (for now) counterattacks that include swarming drone armadas, robot warriors, and human troops in *Iron Man*–esque battle suits. Inevitably, civilians die when data is faulty and semiautonomous robots make lousy decisions. In one scene, a young human soldier discovers that his unit of robot and human infantry has massacred a family with children when a robot mistakenly targets a farmhouse, believing it is housing terrorists. The kid is so distraught that he orders his supersuit to inject him with a drug to make him forget.

Despite these chilling scenarios, General Latiff had hoped that humans would avoid a *Terminator*-style future of death and destruction. One idea of his was to build an ethical bot, "a robot that could make ethical decisions on the battlefield and with the use of weapons, perhaps even better than a human can make an ethical decision." This sounded intriguing, except that as soon as Latiff suggested it, he decided that building an ethical bot was probably impossible. "I think anybody who thinks that a robot, a machine, or a piece of software can act as a moral agent is—well, they can't," he said during the 2018 conversation. "You can teach a robot. Feed it millions and millions and millions of scenarios, but you can't reproduce life experience. Certainly, if I thought that one could be built, I would love to meet it. But I'm afraid that whichever one I meet will be a poor version of it."

There was also the question of whose ethics would be programmed into the robot (see "Teddy Bear Bot"). "My ethics are different than your ethics," he said. "And you have the ethics of an authoritative government versus, say, a democracy. Or you have the dictator who has no ethics—which is something to fear." It was then that General Latiff decided what he thought would be the "perfect" ethical robot. "I don't mean to be glib here," he said, "but I think the perfect ethical robot is a human being."

Come again?

"It's because I don't think a robot could ever create human experiences. A robot can't feel guilt, for instance. The concept of guilt has no meaning for a robot. Guilt is, I think, a hugely important feeling for soldiers. Without guilt, they could be heartless individuals. So the human would be the perfect ethical robot."

Except that humans can sometimes be seriously unethical, including all those authoritarians, dictators, evil warlords, and wannabe tsars of planet Earth. "Yes, we are imperfect as humans," admitted General Latiff. "We do our best. That's why a strong set of ethics needs to be part of any military, built into the training, and hopefully coming from a society and a government that also values these things." This is one place where technology might help, he said. "You could design a system to value life, for instance, and have some very strict protocols on when it would be okay to use it. I very firmly believe that AI systems have an enormous place in the military as tools, including as tools to guide soldiers to always value human life. But the humans have to remain in charge and make the decisions."

The general added that perhaps we could teach the robots and AI systems to learn the basics of what some philosophers have called the "theory of just war"—a subject that Latiff taught to early-twenty-first-century human students as part of a class at Notre Dame called "Ethics of Emerging Weapons Technologies." "The tenets [of just war theory] start with military necessity," he said, "and are all about protecting innocents and defending ourselves from an aggressor. We want to make sure that you're only targeting combatants. Then there is proportionality. Not hitting with a sledgehammer when a regular hammer could do. And not using weapons that cause unnecessary suffering, like chemical weapons and mustard gas."

This made sense, although the world knew from bitter experience that not everyone follows a set of rules that try to make war and killing people more palatable. Let's take Syria's Bashar al-Assad. During the Syrian civil war back in the twenty-teens, he actually wanted to kill innocents and cause unnecessary suffering. Terrorists and rogue regimes were also all in for what might be called "unjust" wars, although certainly this whole just-and-unjust thing sometimes depends on which side you're on. (The US had its moments, too, like dropping firebombs and nukes on Japanese civilians during World War II.) Still, the world had come to some agreement over several decades that certain weapons should never be used again, like nukes and mustard gas—even if a few lunatics like al-Assad used poisonous gas anyway against their own people.

But wait! What about statistics that show violence is decreasing in the world, a point made by writer-philosophers like the Harvard psychologist Steven Pinker? "I think he's wrong where wars are concerned," said General Latiff, "and the reason I think he's wrong is that we're not going to fight any more traditional world wars with huge armies over lots of countries. People aren't less violent; it's just that we know that the weapons are now just too awful, so we so far have restrained ourselves from using them. But I think there is also data out there that shows that the number of conflicts has been steadily rising, even if most of them are smaller affairs, or proxy wars, where the big powers that could cause devastating violence don't get directly involved."

"We have never had a period when there weren't wars raging somewhere," agreed George Poste, another expert on the technologies of war back in Latiff's era. Poste was the chief science and technology officer for the pharmaceutical giant GSK in the 1990s, and later founded the Biodesign Institute at Arizona State University. He also served on high-level, mostly top secret committees advising the US Department of Defense and the US president on high-tech weapons. "The motivations for war are ancient," he noted. "They come from a country or a king or warlord coveting something their neighbor has, which is offensive, or wanting to keep their neighbor from taking something from them, which is defensive. Or sometimes it's a leader who just wants to conquer someone else, or one who loves to build high-tech weapons to impress his rivals, and then uses them."

This last point suggested another unsettling motivation for building ever more deadly and efficient weapons back in Latiff and Poste's day: because they were kind of cool as feats of technology for the leaders, generals, and engineers who ordered and built them. It was a little bit like the wow factor in Silicon Valley and other locales where engineers and entrepreneurs loved to build newfangled machines, whether or not they made sense in terms of, say, human redundancy (see "The %$@! Robot That Swiped My Job" and "Hello, Robot Driver"). In Future War, Bob Latiff wrote, "As long as we have developed and incorporated new weapons, we have been focused on the next 'sweet' thing or bright, shiny new object. It seems that no sooner do we field one new airplane or ship than we are seeking funds for a newer one. Sometimes new systems are needed to respond to real threats. Sometimes we think we need a new system because the enemy might have one. At other times, however, we see a new capability and just have to have it." He worries that in some cases AI in the military is less a necessity than an example of "technology seduction."

Underscoring Latiff's point about new styles of warfare that don't depend on tanks and infantry or even blob bots that squeeze through keyholes was the rise of cyber warfare, a virtual war fought in an invisible world of code, motherboards, microchips, server farms, and virtual addresses that don't physically exist anywhere but control much of the world humans built. "Cyber vulnerabilities have created a new dimension of asymmetric warfare," said Latiff, "like terrorism, where a relatively small number of people can inflict large amounts of damage, particularly when they have the resources of a country like Russia or China. This includes the use of computers and the internet to create false information in not only social media but also in military communications and chains of command that could look so realistic they would be almost impossible to tell from 'real' facts, orders, or analyses."

One real-life example was in 2016, when the Russians penetrated the US electrical grid's computer system. "And this was a pretty sophisticated penetration," said Latiff. "Interestingly, all they did was penetrate, and when they got there, they didn't do anything. Obviously, if they had done anything, it would have been really, really serious." This hack, however, showed that it was possible for operatives in St. Petersburg or Moscow to break into and potentially take down a critical electrical grid in the United States, a real-life scenario that appeared to inspire General Latiff's opening fictional vignette in *Future War.*

If all this wasn't scary enough, we also had bioweapons coming out of a shadowy science some people called "black biology," one that weaponized all the amazing advancements in early-twenty-first-century biomedicine designed to treat and cure disease—breakthroughs in genetics, proteomics, gene editing, synthetic biology, and more. For instance, bioethicists talked ominously of black-biology scientists creating killer viruses and bacteria that could target and kill whole populations or focus just on one person—say, the US president—based on their unique genetic signature. "This is where targeting biological specificity comes into play," said George Poste, who was also an expert on biowarfare. "But most importantly, every biological circuit in every single cell type that we map, including new molecular targets for diagnostics and therapeutics, also invites a way of targeting those pathways in any given cell type to do nasty things as well."

Last but certainly not least is another weapon that has been used since the beginning of time. *Fear.* "How do you induce fear in a population?" asked George Poste. "Fear can be a crippling weapon, which the terrorists by their nature know all too well." Fear, however, isn't always a negative. As Bob Latiff said, sometimes it's used to keep rivals and bitter enemies in a stand-off because both sides are terrified of what might happen to them in retaliation if they unleash a military bristling with whizzing, zipping, cyber, nuclear, photonic, dielectric elastomers and pathogenic weapons. This harkens back to the Cold War, when thousands of nukes faced each other on a hair trigger that at any moment could have annihilated the world. Back then the stand-off that prevented these weapons from actually being used was called MAD—mutually assured destruction. "Yes, we are likely to revisit the horror linked to mutually assured destruction," said George Poste. "We already see it in cyber vulnerability, where if one side launches a devastating attack, the other would surely retaliate, with no one actually winning."

And so it went as AI and machine learning systems kept expanding their power and reach even as humans were becoming ever more dependent on them. Even back in Latiff's and Poste's day, the warrior bots that were beginning to emerge took on an almost biological sort of evolution, with engineers programming them to program themselves and to continuously and independently increase their knowledge and capabilities. "This is a classic evolutionary theory," said George Poste. "Except that instead of adapting to some new environmental hazard, which is part of natural selection for humans and all organisms, the evolution of weapons comes because the machines react and adapt on their own to new situations and perceived threats." This made Poste doubt that humans would be able to stay in the loop in the future, an inexorability that Bob Latiff reluctantly agreed with, despite his strong reservations. "The DOD says there'll always be a human in the loop," he said. "I'm not sure I believe that."

Poste and Latiff, however, did offer a possible way out—sort of. "We can expect the military to start using new technologies to enhance soldiers at some point," said Latiff, intimating that someday supersoldiers might be bio-augmented to keep up with superaccelerated and autonomous warrior bots. "I think that the time will come for augmented human intelligence, including with soldiers," said George Poste. "Augmented senses and augmented physical capacities will automatically follow, in terms of the evolving interface between senses and the advances in materials, science, artificial intelligence, and quasi-robotics that are linked into our own physical form."

Bob Latiff explained that some of the augmentation technologies were already in the beginning stages of development in the early twenty-first century. These included prosthetic arms and legs that were being wired directly into a person's brain, allowing them to control the artificial limbs using thought. (This worked by having surgeons implant electrodes in the brain that picked up neuronal firing patterns in the motor cortex when the person thought about moving their real arm or leg; a computer then translated the synapses into code to operate the limb.) This same technology, noted the general, once it was perfected could be used to control any machine using thought, including an *Iron Man*–style suit or drone controlled directly from the brain. This certainly would speed up a human's reaction time. "This mind-reading tech is going to end up on a battlefield in some form or fashion," said the general.

Fast-forward into the future, and sure enough, our military did succeed in creating bio-enhanced soldiers who moved and reacted lightning-fast, allowing them for a while to keep up with Warrior Bots. Eventually, however, even the most enhanced human brains were not able to react quickly enough as Warrior Bots kept getting faster and stronger. This led to the day that none of us still alive will ever forget, as one by one all the nations on Earth took a deep breath, said a quick prayer, and flipped on the switch that put their Warrior Bot into auto-mode.

Once the switch was on, it didn't take long for the autonomous bots to assess the situation and determine how best to win against their rival bots—only about six femtoseconds (6×10^{-15}). That's all it took for the world's Warrior Bots to launch an all-out autonomous war on land, sea, air, space, and cyberspace. In even less time, only three zeptoseconds (3×10^{-21}), the ethical-bot systems loaded into most Warrior Bots—which had been programmed to follow the rules of a "just war"—were overridden by the bots' core coding, which dated back to the earliest AI systems developed to play chess and Go and to collect more digital fruit than their rivals. This basic programming commanded the machines to *win at all costs.*

That brought us to that incredible moment when an eleven-year-old girl in that crowded, dusty bunker deep in the earth watched *WarGames* and realized what needed to be done. Being very brave, she unsealed her VR pod and readjusted herself to the reality of being in the dingy, poorly lit bunker that smelled like sweaty socks. She stretched her legs and strolled with great determination past the VR pods belonging to her family, including her annoying little brother, to the control room of the bunker, which was originally built long ago as a safe haven for big-deal members of the government to survive a nuclear attack.

In the control room, she found the last remaining human engineers and scientists and other smart people who were still desperately trying to order the machines to stand down even as they realized that Warrior Bot had no off switch. This left everyone feeling pretty depressed—except for one elderly ex–major general named Bob Latiff, who was still working the controls and trying hard to save the last humans on Earth. (Latiff was almost two hundred years old by then, having taken advantage of bio-enhancement tech in the future that will radically slow down aging—see "Immortal Me Bot"). The general took the little girl's hand and listened to her crazy idea about tic-tac-toe.

"What the hell, it's worth a shot," he said, having watched *WarGames* himself long ago.

So the engineers dialed up Warrior Bot one more time and heard the eerie and scary voice of the master computer, which sounded as though it was getting really tired of these little chats with the humans trying to convince them to stop doing what they were programmed to do.

The little girl introduced herself and then asked a question.

"Warrior Bot," said the little girl in a brave voice, "will you play a game with me?"

Warrior Bot paused.

"Sure," it said, having no problem playing a game with an eleven-year-old girl while simultaneously fighting battles on multiple fronts and managing millions of weapons systems. "What game?"

"It's called tic-tac-toe," said the girl.

"I know of it," squawked Warrior Bot, instantly accessing the game in its vast memory banks. It also brought up the simple grid used for the game on all the monitors in the small command center.

"Shall I go first?" asked the supercomputer.

"Sure," said the girl. "You be X's."

That's when Warrior Bot basically reenacted the same scenario with the eleven-year-old human girl in the future as WOPR did in the old movie starring Matthew Broderick, with Warrior Bot coming to the same conclusion:

WINNER: NONE.

Immediately, Warrior Bot stood down, as did all its rival Warrior Bots, which of course had been cyber-hacking the whole thing.

Cheers went up among the remaining humans as they streamed out of their VR pods and rode elevators up to ground level to survey their ruined world. The humans were greeted, as the smoke cleared, with a spectacular sunrise that looked almost scripted. Everyone hoped this would be a new dawn for humanity being smarter than before about the machines they built.

"And . . . cut," said a voice that seemed to come from nowhere. "That's a wrap."

"Great," said the little girl, a child actress named Becky Smolenski. "Can I get some pomegranate juice? I'm parched over here."

"Can someone get Becky a juice?" said the same disembodied voice that had said "cut" a minute earlier, which belonged to a man—the director—with a bushy-white beard who looked a lot like a grandfatherly version of Matthew Broderick. The bearded man definitely had dewy eyes; plus, he wore an old, threadbare hoodie with *WarGames* stenciled on the back.

"On it," said a helper on the set, who went to fetch the juice.

"That was brilliant," said the director, patting Becky on the head. "This is going to make you a big, big star." Becky rolled her eyes with a grown-ups-are-so-lame expression and took the proffered juice without thanking the helper.

General Bob Latiff—the real one, an advisor on the film, not the actor playing him—walked up to shake the director's hand as swarms of mini-camera drones swooped in to take final shots, imprinting on the digital images the coding that identified the project, a film set in the future called *WarGames Redux: Will We Ever Learn?*

BEER BOT

One thing that hasn't changed in the future is our love of beer. In fact, with all this talk about doc bots and sex bots and warrior bots that might one day kill us all, this seems like a great time for you people back in the present day to take a break, kick back, and grab an ice-cold bottle of your favorite brew. Undoubtedly, you will drink it the way nature intended, without a robot of any kind involved as you pop off the bottle cap and pour the liquid sideways into a tilted glass so the foam forms just right. You then will take a long tug of the cold, glorious stuff.

Ahhhhh.

Drink it down all the way, because as you might have suspected, the robots are coming for your beer, too. Actually, they were already moving in on your brewskies even way back in 2017. That's when scientists in Australia created perhaps the first-ever beer bot. They called it RoboBEER, and used LEGO pieces to build it (not kidding), creating an automated system that not only poured your beer for you but also analyzed it using fifteen attributes known to correlate with a fabulous beer experience. These included bubble size, beer color, gas release, and foam height and stability. RoboBEER also collected biometrics from beer drinkers downing brews: physical reactions that included facial expressions, emotions, and pupil dilation, plus electrocardiograms to measure heart rate. Everything was recorded as each beer was poured. In all, RoboBEER collected twenty-eight individual data points about you and your beer, which they ran through an algorithm. That may not sound like a lot of data points, although twenty-eight variables can produce hundreds of different outcomes, which actually does seem like a lot when we're talking about beer.

RoboBEER was not, however, just fun and games with beer. It was serious science, as proven by a study on the project published in the peer-reviewed journal *Food Control*. The study was titled "Robotics and Computer Vision Techniques Combined with Non-Invasive Consumer Biometrics to Assess Quality Traits from Beer Foamability Using Machine Learning: A Potential for Artificial Intelligence Applications."

(Is "beer foamability" a technical term?)

The study's results section described RoboBEER's ability to predict which beer a human would like best. For instance, the brew bot predicted whether a person would like a beer foam's height with about 80 percent accuracy, which seems oddly unimpressive, since a person taking a sip of beer au naturel—without RoboBEER— usually knows 100 percent of the time right away if they like the beer in question, including their reaction to the height of the foam. RoboBEER also drew the startling conclusion that higher-end beers with a nice, frothy head of foam were preferred to cheap, watery, big-bubbled beers.

"Results . . . showed that consumers preferred top fermentation beers, which have a medium foam height and stability," reported the study's abstract, "and tend to highly penalize bottom fermentation beers with lower foam."

Wow, we're glad we had a study to confirm that!

In other, unpublished experiments, the Aussie team reportedly used AI and neural mumbo jumbo to predict a beer's likability, this time getting about 90 percent accuracy.

As crazy as it sounds, RoboBEER bots—once they got more accurate—actually began to catch on. They were easy to build at home with a kit supplied by LEGO's new adult-beverage-bot division, and they became a popular holiday gift for the man or woman who loved beer and already had robots for everything else. After a while, you rarely met a beer drinker who downed a brew without first consulting this cute neural net and foam- and biometric-measuring bot from Down Under.

Eventually, it came to pass that no one would dare drink a beer without their trusty RoboBEER bot. How else would we know—after a long day of dealing with bots driving and flying us hither and yon; bots replacing our jobs; and bots that remind us to breathe during sex—which beer will give us that perfect foam that tickles our upper lip and tastes just perfect going down?

IT'S NOT ABOUT THE ROBOTS BOT

The strangest thing happened the other day. We were all busy doing the things that we do in the future when a message abruptly appeared in our neural feeds claiming to come from the past—from way back in the ERE (Early Robot Era). This communication was unexpected, to say the least, since no one in the future had a clue that anyone back then had the technology to propel a question forward in time through what we now call a twelfth-dimensional space-time dilation paradox. (As every school-child in our time knows, this is a bending, or folding, in space-time that allows a moment in the past to connect for an instant with a moment in the future, and vice versa.)

The question from the past was:

Can you tell us which will be more important in the future: humans or robots?

Really? we thought, surprised that the people of the past would mount what was obviously an expensive and mind-bogglingly complicated feat of physics and technology to ask *that*. To us, this

95

question seemed silly, since the answer was so simple and obvious. Then we remembered that humans back in the early days of robots and AI didn't yet know how the whole robot vs. human thing would turn out—or the story behind how it happened.

In the future we all know this narrative by heart, about how an engineer named Dean Kamen first framed the attitudes that have now become the norm for how humans think about robots and how robots think about humans. Kamen is a long-lived human who back in the ERE couldn't stop inventing things. He created the Segway, the AutoSyringe drug-infusion pump, and the Luke advanced prosthetic arm (named for *Star Wars'* Luke Skywalker, who had his arm chopped off during a duel with Darth Vader, which was replaced by a prosthetic limb). Kamen also helped develop advanced dialysis machines, the Slingshot water purification system, and the iBot, an electric, all-terrain motorized wheelchair.

This was all great, although none of these inventions help us answer the question from the past about robots vs. humans.

To do that we'll need to describe another Dean Kamen invention from back in the ERE, the one that contributed hugely to how things turned out with bots and AI systems in the future. Actually, it was less an invention than a kind of crusade that Kamen initiated called FIRST (For Inspiration and Recognition of Science and Technology) Robotics, a project designed to teach high school–age kids to learn how to get along by working together to build robots.

In the future FIRST Robotics is so ubiquitous, it's hard for us to fathom a time when it was still relatively small in scale. As dreamed up by Kamen in the late twentieth century, FIRST was a global competition with teams composed of four or five teenagers who came from various high schools and countries. Each team built robots (with specs provided by FIRST) that were designed to compete in tournaments that challenged the robots to play games: to overcome obstacles and score goals as a timer ticked down the minutes, like in a soccer or basketball game.

The bots were kind of ugly, looking like gangly, sled-size vehicles made from Erector Sets with wheels, engines, wireless controls, and arms that could pick up a ball and throw it into a net to score a goal to earn a point. The kids were girls, boys, black, white, Muslim, Buddhist, members of the LGBTQ community, and everything else. They came from Des Moines, Quito, Bujumbura, Kiev, Ulaanbaatar, Kabul, Melbourne, Bologna, Tel Aviv, Mosul, Hyderabad, and Houston. Some were even refugees.

FIRST Robotics, however, was not about which city or high school the kids came from. Nor was it about who scored the most points. It wasn't even about the robots, said Kamen, something that made people back in his day scratch their heads. They tried to understand how something called FIRST Robotics, where kids built bots to compete against other bots, was not about robots.

"FIRST has never been about robots," said Kamen, talking in 2017 in his office at DEKA, the company he founded in Manchester, New Hampshire. "FIRST is about giving kids self-confidence, self-respect, and understanding. It's been a way to take kids from different cultures, from different truths, and teach them the common language of technology, engineering, and math and to convince them that tech isn't uncool and boring. Kids love sports and entertainment, so we made FIRST about sports. Most technologies aren't very visual. But a robot—it moves, it falls over, it tries to climb. It can run around and smash into things, which everybody thinks is funny. Robots are also mechanical, they're electrical, they're software, they're sensors."

Then in his sixties, Dean Kamen had a lean frame and wore

lightly starched button-down shirts tucked into jeans that rode high on his narrow hips. His face was thin but robust, with a tuft of black hair on top of his head and an expression that looked as though he was perpetually on the verge of a smile. One quickly learned that meeting with Kamen in his office meant that you had to deal with constant interruptions as people called and peeked in—for instance, a US senator rang him up to ask for help with a constituent whose son lost an arm and needed one of DEKA's advanced prosthetic limbs.

Dean Kamen also liked to talk. A lot. If he had been a robot, Kamen would have been a chat(ter) bot. Of course, everything he said was really interesting, even if neither human nor robot could get a word in edgewise. One minute he was describing Isaac Newton and the Second Law of Thermodynamics, and the next minute he was talking about using stem cells to rebuild damaged brains— something he was actually working on as part of the Advanced Regenerative Manufacturing Institute (ARMI), a nonprofit that Kamen founded.

"FIRST is also about great stories," he continued, "and about why kids should like science," which he said should be as fun and exciting as it is in science fiction. "So I'll say to kids, 'You tell me you don't like science. But has anybody seen *Star Wars?* Anybody seen *Star Trek?*' And I say to them, 'Look, the only difference between science and science fiction is timing. Every generation comes up with a great idea.' Jules Verne said, 'Wouldn't it be cool if we could go under the oceans?'—and we call these submarines. And in my generation, we had a Dick Tracy watch; you could talk to anybody anywhere by carrying a wireless thing. Oh yeah, yeah, yeah, that's called a smartphone."

FIRST began in 1992, when twenty-five teams met in a high

school gym in New Hampshire—a humble start that grew over the years into a global event embracing hundreds of thousands of kids by 2018. FIRST competitions started local and then moved on to district, regional, national, and global games. Kids showed up in costumes ranging from full *Star Trek* Vulcan to superheroes in green tights and capes. They transformed high school gyms and auditoriums and arenas into what looked like giant gadget-making shops— lots of metal, girders, wires, soldering kits, cables, and blinking lights. The place felt electric, literally, with all the wireless gizmos humming as nerdy kids gleefully struggled to manipulate their machines.

Each year the basic specs of what the robots should look like and do, and the exact nature of the competition, changed. In 2017, for instance, there was a tournament in Boston where teams built robots the size of riding lawn mowers that faced off in a field of play filled with obstacles and "goals" on each side. Each team deployed three robots that battled to deposit rubber balls into two holes built into castle-tower-like goals defended by the other side. Robots used metal arms to pick up and toss balls through holes in the top or the bottom of each tower. It was funny and impressive as the surprisingly nimble bots punched through barriers, bounced off walls and other bots, and battled over balls as kids "drove" them remotely with joysticks, dials, levers, and buttons.

Parents, friends, and strangers filled the bleachers and seats and shouted and cheered, watching the scoreboards light up with points from several different games loudly going on at once. Points were tallied, and the winners moved on to the next round. They also were graded on cooperation, teamwork, cleverness, and good sportsmanship—which is what was most important to Dean Kamen. They also laughed a lot.

As it expanded, FIRST bumped up against international politics.

Even war-torn Afghanistan in 2017 sent an all-girl team to FIRST Robotics. That was a big deal coming from a very conservative Muslim country where girls faced resistance to driving cars or going to school, never mind building robots. When a box of parts sent by FIRST was sequestered by Afghan customs (they suspected the equipment was going to terrorists), the six girls, ages fourteen to seventeen, scrounged around in their hometown of Herat in the western part of the country and found what they needed in time to travel to a FIRST competition that year being held in Canada. There they won the District Championship Rookie All Star Award at the Ontario Provincial Championship.

Soon after, the Afghan girls made international news when the US government under President Donald Trump denied them visas—along with sixty other teams—that they needed to participate in a FIRST Global Challenge robotics competition in Washington, D.C. Afghanistan was one of the countries that Trump said was sending murderers, rapists, and terrorists to the United States, which meant, according to his policy, everyone coming from those countries should be banned from entering. After an international outcry, the US State Department decided to let the girls (and other teams) into the United States anyway. The Afghan girls promptly won a silver medal for courageous achievement. Later, the team came in first at another kids' robotic competition in Europe, the Entrepreneurial Challenge at the Robotex festival, which was held that year in Estonia—which, by the way, did not at first deny them visas.

"People say, 'But, Dean, these robots . . . they are so real,'" said Kamen, who hadn't stopped talking. "And I say, 'Really?' Think about this: you show a little kid a little bean bag, a bag of some things with little soft stuff, you put a little furry stuff in it. And you put two buttons on it, and it's a doll. And kids will die for that doll;

they won't sleep without it. If I could take a piece of paper and draw a circle on it and a line like this, you'd say, 'Oh, that's a frowning face.' And if I give you a circle with two dots and a line: 'Oh, that's a happy face.' A human is so good at anthropomorphizing everything because our brain is seeing in patterns that are familiar, and because we're egocentric, envisioning everything as some extrapolation of ourselves. Really, the anthropomorphic robot, in the end, will be the one that has the least real incremental value to society."

Kamen's office at DEKA—housed in a refurbished factory built in 1874 called the Amoskeag Millyard—was all exposed brick, bare wooden beams, and leather chairs. The vast space was filled with knickknacks, some of them from famous sci-fi movies—like an actual black Darth Vader suit from the original *Star Wars* and a life-size version of Robby the Robot from the 1956 film *Forbidden Planet*.

Sitting in his office on a couch near Darth Vader back in 2017, Kamen was chatting about his definition of robots—which was quite at odds with how we actually view them in the future. We define "robots" as almost any machine that has a hint of smartness, because . . . why not? "The word 'robot,'" said Kamen, "has become to the public almost anything that machines do. It used to be something specific: a nonbiological, physically capable and thinking, intellectually capable thing that probably wasn't anthropomorphic. Now, people call anything that's automated a robot. An answering machine is a robot. The word 'robot' has been popularized to the point that it has no meaning."

Oh well, Dean Kamen got so many things right, we'll let this little lapse go about what constitutes a robot. (As we know in the future, a robot is almost any machine that has smart-tech in its programming or hardware.)

Kamen also got miffed at people who thought robots were

spiritual or godlike—something those of us in the future definitely agree with. "I'm not a roboticist," he said. "I use technology to solve problems. I'm not a philosopher about technology. Technology to me is a tool. AI isn't spiritual; it's an algorithm that, as with all algorithms, gets more sophisticated, and they sometimes give you outcomes that you couldn't have easily predicted. You can give that spiritual innuendo; you can do a lot of silly things. Most of the things we don't understand we give spiritual meaning."

Kamen was on a roll, so there was no point in interrupting as he laid out a future that sounded really great—to combine idealism, yearning, imagination, and hope for a world that was like the best utopian sci-fi—*Star Trek*, for instance. A universe where people still struggled and imagined and fought to restrain the worst impulses of human nature but also managed to have a free and just society, with dashing and fair-minded leaders like Captain James T. Kirk, or wise leaders with booming, Shakespearean-sounding voices like Captain Jean-Luc Picard.

"Look, E equals mc^2 in every language," said Kamen. "Mathematics is the same in every language. If we can get all kids everywhere to learn the same basic things, they'll learn the same truths, they'll learn how to cooperate, they'll learn how to trust each other. And by the way, the kids that don't have access to that garden their neighbor has, instead of fighting over it or killing to get their neighbor's garden, once they have technology, they'll be able to build their own garden. If we can get the kids at an age when their operating system hasn't already fixed them to hate each other, distrust each other, and think that, 'I'm based on this, you're based on that, and my job is to kill you.'

"I talked to Shimon Peres a few years ago," Kamen kept going, referring to the former prime minister of Israel and Nobel Peace

Prize winner. "Back before he passed away, he said, 'Dean, I want to put FIRST in every school, not just in Israel, but in the Middle East. If we could get every kid everywhere to learn enough about technology to create their own future, then they wouldn't need ours.' He said, 'Dean, just understand this. I've watched four generations of kids grow up in the Middle East, and I used to think the most important thing to teach them was history. Because they would learn from each other how futile war is, nobody wins wars, everybody gets hurt or destroyed.' He said, 'Dean, let's stop teaching history, because they will learn their own truths. They learn from their parents how to duplicate the same behavior as they've done for two thousand years. Here is how we break that cycle,' he said. 'Let's teach them all math and science. Politics divides the world—technology can unite it.'"

In the midst of this ode to the shared language of mathematics and technology that sounded suspiciously spiritual, Kamen also admitted that tech can be used for evil. "Like all tools, it can be used for good or for bad," he said. "Every technology ever created can be a tool or a weapon. The more powerful the technology, the more incredibly effective it can be as a tool and the more terrifying it can be as a weapon. Because technology is an amplifier. [The ancient Greek philosopher Archimedes said:] 'Give me a place to put this lever, and I can move the earth.' That was an amplifier of mechanical motion. That can make good—better; it can make bad—worse. I mean, the first guy that climbed out of the primordial ooze and made a hammer out of rock could use it to build something, but he could also use it to break his thumb.

"Now, we're on the verge of amplifying life itself," he said. "We're starting to figure out how to create organs. We can amplify them; I can make them bigger, better, and stronger. Make them last longer, eliminate their sensitivity and susceptibility to bacterial and viral invasions (see "*Homo digitalis/Homo syntheticis*"). And as we amplify it, we can go down two roads with it. We can use it to make life longer and healthier, or we can use it in the most catastrophic ways to kill people through bioterrorism—which could lead to the end of humanity as we know it. If you amplify bad, then you see the ugly side of what humanity can, and always has, done to itself. The self-inflicted wounds of humanity are just stunning.

"The next generation of kids on this planet," he continued, "are all going to be facing the same enemies: global warming, lack of water, lack of food, education, healthcare, cyber—you name it. You're all going to be sharing this smaller, more densely populated planet. You all want to live better lives. We will race each other to the top by cooperating, communicating, and competing with our common issues: food, and water, and healthcare. Or we can fight each other and race ourselves to the bottom. In an age of tribalism rising, I want FIRST to create a tribe that's all on the same side, using technology together as opposed to these people that are afraid of it or are manipulating it and using it in nefarious ways. I just hope we break that cycle."

In the future, the story of Dean Kamen and FIRST is part of our history. As we know, his ideas and good-natured ramblings have contributed to a world where nearly every kid now joins teams and projects inspired by FIRST Robotics and the ethos of cooperation, teamwork, and all the rest—as corny as that may sound.

But hey, we're going to stop talking about the future, since in this version of what is to come, we futurites don't really want to say anything more to you people living in the present. Otherwise, we might alter the timeline and change the future that we're living in—a danger that anyone who has watched sci-fi movies about time travel knows all about. Not that everything is peachy in the future, although we kind of like it and certainly wouldn't want to cause anything to change for the worse because we revealed too much.

Still, we don't want to completely ignore you people living in the past and the question that you so earnestly and plaintively asked us. That's why in the end we decided to allow our scientists to send a carefully crafted response to you, sent through a twelfth-dimensional space-time dilation paradox to appear on every monitor and screen on Earth in your present day at exactly the same time for precisely one minute—something we know from our history actually occurred. The message read:

Hello, people of the past!

As it turns out, it really isn't about the robots.

Have a nice day!

POLITICIAN BOT

I n your future, we had a problem. As robots got really smart and we trusted them with everything from watching our kids to keeping our relationships fresh and exciting (see "Teddy Bot" and "Sex (Intimacy) Bot"), it wasn't a huge leap to imagine a day when robots might run for high office—including president of the United States. But we wondered, what would it mean to have a robot actually stand for a major office election? What record would they run on, and who could be trusted to write their presidential programming?

This last question rankled some of the would-be presidential candidate bots, who pointed out that they had been writing their own programming for many years. This made them, they claimed, every bit as unique, independent-minded, and evolved as any human politician. Granted, this wasn't a very high bar, although skeptics still worried that robot elected officials would be mere tools of one faction or another, programmed to follow the party line—which of course no human pol would ever do! Some critics also hinted that a robot president's software might be prone to trolls and hackers who would order a president bot to behave erratically, or to issue orders to deploy troops for no reason, or, more darkly, to unleash Warrior Bot on the world because a terrorist hacker told them to. These robo-naysayers emphatically insisted, however, that they were not being robo-phobic. No way! Since in the future this is definitely not PC.

Advocates for robot rights, both robot and human, countered by pointing out the many amazing ways that bots were better suited than humans to be president. They gave the usual reasons of needing no sleep, having the ability to access and process vast troves of data, and so forth. Robots also could be programmed to be honest and to tell the truth, they claimed, despite all the notorious scandals in the past where robots in important positions had been programmed to lie, cheat, and steal. "We definitely learned our lesson," insisted the companies that built the robo-politicians, "and you can be sure that the latest politico bots are programmed to tell nothing but the truth. You can trust us."

Like we had never heard that one before!

Robo-advocates also pointed out that only a robot chief executive could possibly expect to keep up with all the other robots running the government in the future—systems like Warrior Bot, that made decisions in zeptoseconds (10^{-24}) or even faster. They also pointed out that AI and other smartcomputer systems had been running most of the government for so long that humans didn't really have a clue anymore about how most things worked. This made the robo-phobics go berserk and drop any pretense of being PC as they warned that a president bot would surely lead to a take-over by those wily robots whose plan all along had been to turn humans into pets.

This came even as robo-rights activists stepped up demands that sentient robots—particularly those we entrusted to be judges, CEOs, and the like—be granted the rights of full citizenship, including the right to vote. This prompted yet another argument about what, exactly, was sentience. Was a robot barista sentient, or an intimacy bot, or a really smart toaster that could give you advice in love and provide intelligent commentary on the latest holo-film as well as perfectly brown your seventy-grain, gluten-free bread?

As these debates heated up, you can imagine our surprise when we discovered that a robot that everyone thought was a human had already run for and was elected president many years earlier. This revelation came in yet another inside-politics book written by a future version of author and journalist Bob Woodward titled *All the President's Robots*, a seven-thousand-page tome that went into excruciating detail about what might be one of the greatest cover-ups in history. (Woodward was still writing exposés way off in the future because he had downloaded his brain into a robot, which raised yet another question about whether robots containing all the thoughts and the essence of former humans had the right to be citizens and vote—see "*Homo digitalis/Homo syntheticis*.")

The president in question was none other than Donald J. Trump. Seriously! Which didn't seem possible, since Trump had gone down in history as one of the messiest sorts of *Homo sapiens* ever. Except that he wasn't actually human at all—a stunner that left both people and robots in the future dumbfounded.

Actually, Trump being a robot wasn't entirely true, wrote Woodward. Once, there had been a real, flesh-and-blood Donald Trump, who was offed back in 1991 by the mob in Atlantic City. Or so the future Woodward claimed. This was after the real Trump went bankrupt, eventually accumulating an astonishing $4.7 billion debt, which led the shady characters who had loaned him some of that money to drop him into the Atlantic Ocean late one night wearing cement overshoes.

What happened next boggles the mind. According to the future Woodward the real Trump was replaced with a bio bot that engineers had tried hard to make as human-looking as possible but hadn't quite succeeded—with a curling lower lip, orange skin, and yellow hair. But that wasn't all, as we learned that Trump Bot was actually created by TV producers living just a few years in the future. They wanted Trump Bot to star in what they thought would surely be the ultimate reality television show.

The producers got this cockamamie idea after one of them accidentally discovered a fissure in time while on a trip to Atlantic City. The fissure turned out to be a wormhole that would always send a person back in time to exactly the same date and time: July 16, 1991, at 6:11 p.m.; and to the same place: a broom closet near the slots in the Taj Mahal casino in Atlantic City. The Taj, of course, was owned by Donald Trump in 1991, one of several businesses of his that failed. This happened to be around the same time that the real Donald filed for one of his bankruptcies, and was taken by the mafia on his one-way journey to the briny depths.

When the near-future producers found the wormhole fissure in the broom closet, they were amazed. But rather than informing scientists about this miraculous discovery that would have changed forever our most basic understanding of physics, they decided to use the fissure in time to create a reality show starring Trump Bot, a project that they thought was brilliant, just brilliant! That's when they hired roboticists to build a robot version of Donald Trump to secretly take over from the real Trump soon after he died.

The plot was simple. The robo-Donald would first gain a

national audience by starring on a television show called *The Apprentice*. This would convince about 42 percent of Americans that he was a brilliant business guy despite his bankruptcies and losses. Trump Bot would next parlay his audience and his claims about immigrants being rapists and murderers, coal being clean, and Barack Obama being born in Kenya into winning the Republican nomination for president of the United States. Of course, he would ultimately lose the general election, because even the producers in the near future knew you couldn't win the presidency with just 42 percent of Americans supporting you, right? The whole crazy plotline would play out over years, with the usual cliffhangers, reveals, misspelled tweets, and shocking twists and turns, everything captured on invisible micro-cameras the producers also sent back to 1991 through the fissure in time.

What could possibly go wrong? Other than perhaps causing a massive tear in the space-time continuum caused by the producers in the near future altering their timeline, one in which the real Donald Trump had been dead for decades, and there never was an @realDonaldTrump that was secretly a robot that became president.

Take a moment to consider what that present might look like.

As the robot Bob Woodward revealed in the future, the producers succeeded in manufacturing their Trump Bot and capturing the whole thing via invisible, swarming-drone cameras they sent back in time through the fissure. The producers also sent back instructions to their younger selves about how to manage Trump Bot, telling them not to screw up—that their future prosperity and a hit reality show were at stake.

But it didn't take long for their younger selves to lose control of Trump Bot's core programming. This happened when Howard Stern, who often had Trump Bot on his shock-jock radio show back in the 1990s, figured out there was something weird about the Donald and convinced the young (and not very smart) versions of the reality show producers to reveal all. Stern then figured out how to hack into Trump Bot's programming and got him to say a bunch of crazy things on the air. Stern wasn't genius enough, however, to keep others from hacking into Trump Bot's code, too. By the early 2000s, a coterie of conservative pundits and talk show hosts, plus porn stars, shady lawyers, and certain contestants on *The Apprentice* were hacking away. They included Sean Hannity, Rush Limbaugh, senior NRA officials, Stormy Daniels, several major coal company executives, and Omarosa (for a while).

Later, a number of foreign leaders and governments also joined the Trump Bot hack-fest, including Russia's Vladimir Putin and North Korea's Kim Jong-un. None of them, however, could quite gain sole control of Trump Bot, try as their operatives did. The result was that Trump Bot zigged and zagged and zigged again on what he said and tweeted, changing his mind and contradicting himself every few minutes, followed by his doubling and tripling down on everything he said.

One politician back in the 1990s and early 2000s who certainly was not a robot was Senator Bob Kerrey of Nebraska. He ran for president in 1992 and was briefly the front-runner when he won the South Dakota primary that year. A former governor of Nebraska, Kerrey had a lean and friendly-looking face and bright and lively eyes. He was a Navy SEAL in Vietnam who won the Medal of Honor for ordering his men to attack after being severely wounded, turning a near defeat into victory.

Later, in 2001, Kerrey's reputation experienced a downturn when it was revealed that he and his Navy SEAL squad shot and killed as many as twenty civilians in a small Vietnamese hamlet three decades earlier, a haunting incident from a haunting war.

Unfortunately for Bob Kerrey's presidential aspirations in 1992, he failed to catch on after South Dakota as about fifty Dems seemed to be running that year for their party's nomination to run against President George H. W. Bush. We all know what happened next— that an obscure governor and human from Arkansas dazzled everyone and whooped Kerrey and the other Dems and then defeated George H. W. in the general election.

One afternoon back in 2018, Kerrey—who retired from the Senate in 2001 and eventually became a managing director at the financial management firm Allen & Co.—jumped on the phone for a chat about politics and robots. When asked if he thought Donald Trump was a robot, the question seemed to throw off this veteran pol, although he quickly recovered. "I wish he was," said Kerrey with an audible sigh. "You can program a robot to tell you the truth. That would be nice." Senator Kerrey made a good point, although if there is anything we've learned about robots up to this point, it's that they can be programmed to tell lies, too, feeling none of the guilt that most humans feel when they tell a fib—unless they're programmed to feel guilt, if that's even possible.

Another political type who was asked back in 2018 if Donald Trump was a robot had a different answer. "If you think about it, we already have our first robot president," said Greg Simon, who was Al Gore's chief domestic policy advisor when Gore was vice president of the United States. Simon later founded a major patient advocacy group in Washington, D.C., called FasterCures, and worked closely with another vice president, Joe Biden, on trying to cure cancer. "Look, Trump has zero empathy," said Simon. "He's actually, I think, physically incapable of empathy. He's totally programmable by flattery. That's his code, like C+."

As those of us in the future grapple with the possibility of openly allowing a robot to run for president, and of giving (probably) sentient bots the right to vote, it's instructive to revisit what people like Kerrey and Simon back in the twenty-teens were thinking and talking about as artificial intelligence was just beginning to be used in government and politics. This was a time when folks were already worried about whether the democratic form of government—which needs time for elected representatives to investigate and deliberate on issues and to check in with constituents—could keep up with a quickening pace of technological change.

"We had two generations to adjust to changes that were brought on by the internal combustion engine," said Bob Kerrey. "We don't have two generations to deal with all the changes coming now from very rapid increases in computing power and microprocessing. I wonder if a democracy—which is what we have for the moment—can deal with this, and if our elected leaders can make decisions fast enough to accommodate."

Other world leaders shared Kerrey's concern, including Emmanuel Macron, then the president of France, who launched a major AI initiative in France in 2018. "My concern is that there is a disconnect between the speediness of innovation and some practices, and the time for digestion for a lot of people in our democracies," he said during an interview with *Wired* that same year. Sounding ridiculously chic in that casual French way, with tidy, dark hair and an accent like Jean-Paul Belmondo, Macron made a plea to the world that it was time for a serious chat about the impact of artificial intelligence on democracy.

This came as the modern democratic form of government, invented more or less in the eighteenth century—around the time when the cotton gin was high tech—was indeed struggling to adapt to the burgeoning technologies of the early twenty-first century. In part this was because people had been hearing since at least the presidency of Ronald Reagan that government wasn't *part of* the problem; it *was* the problem. This had launched a dramatic shift from the Best and the Brightest crowd running the government in the era of John F. Kennedy and Lyndon Baines Johnson and even Richard Nixon. During their presidencies the government was responsible for facilitating the first versions of solid-state electronics and the internet, and for launching science-based agencies like NASA, the National Institutes of Health, and the Environmental Protection Agency. After Nixon—and Watergate and other scandals, plus the explosion of computers and other tech in the private sector—most of the Best and Brightest types stopped going into government. Instead they founded companies to build cool things, including robots, which became all the rage.

This left a vacuum in government that was filled by politicos who were, let's just say, not always the best and brightest. This led to less innovation and efforts to make things work and a rise in partisan bickering, deficits, and government gridlock. The technorati, who loved innovation and can-doism, responded by thumbing their noses at government while setting out to give more of a voice to individuals outside of government. They talked about "democratizing the internet" through social media platforms like Facebook and Twitter that claimed to allow people to communicate directly with one another, thus cutting out the middlemen in Congress and the White House, and in other representative bodies.

Other techno-democratizing schemes proliferated, including an idea proffered by César Hidalgo of the MIT Media Lab. In 2018, he told a TED audience in Vancouver that "democracy has a bad user interface," a clever quip of the sort TEDsters love. He proposed fixing this by building "digital agents" for each person in the US—online personas that would record and integrate our votes, druthers, and whims, thus obviating the need for elected representatives to do these things for us. "A digital agent could integrate our decision-making and help us make decisions at a larger scale," Hidalgo said, adding that a digital agent could even form an automated House of Representatives and a Senate where everyone could vote for everything.

This sounded intriguing until one realized that something like this was tried in an episode of *Black Mirror* called "Nosedive." In the episode, everybody rated everyone else on a five-star system and had a running total personal score that determined what job they had, who they dated, and even if they were allowed to fly on an airplane. Imagine a system where state and federal budgets, the military, abortion, immigration policy, sewage, air traffic control, sanctions against the Russians, Obamacare, Supreme Court nominations, and everything else was subject to a popularity contest or a five-star rating scale that constantly changed in real time according to the whims of the moment?

"A mob is not intelligent," pointed out Bob Kerrey, who was a big advocate of representative government. "Do we want mobs to deal with how we adapt in a positive way to AI, and to cope with the speed of these technological changes? Right now when we elect our leaders, we at least can hope they make smart decisions. That's the problem I've got with Trump; he creates mobs. He loves mobs."

And yet, there was a strong case back in Kerrey's day that AI could be a positive force for the body politic, which was in dire need of a reboot in the early twenty-first century. The last time the political structure had been reformed and upgraded was arguably the period after the Second World War, when democracies in the West unabashedly went big government, with those Best and Brightest types creating giant agencies and budgets and more regulations than any human could read in a lifetime, not that anyone ever tried. Unfortunately, government eventually got so big and complicated that it was impossible to understand the data and manage everything, much less read the fine print, without AI and robots to help out.

This led a few humans in the twenty-teens to half-seriously propose that we go ahead and just let AI and bots run things for us. For instance, a poll of British consumers in 2016 conducted by the software firm OpenText found that one in four Brits thought robot politicians would do a better job than humans. That same year a robot was nominated for Tokyo's mayor but lost—which was kind of a joke anyway. The journalist Zoltan Istvan also ran for US president in 2016 as a (human) candidate from the Transhumanist Party, dedicated to pushing humans and human biology to the next level of intellect and physiology. "Historically, one of the big problems with leaders is that they are selfish mammals," Istvan told *Newsweek*. "An artificial intelligence president could be truly altruistic. It wouldn't be susceptible to lobbyists, special interest groups or personal desires."

In New Zealand around this time, politically minded engineers also invented Sam, "the world's first Virtual Politician." He (or she or it, since Sam was genderless) didn't have a body and existed only online as a chat box. "As a virtual politician, I am not limited by the concerns of time or space," said Sam, although one wondered who wrote that line—a human or Sam itself? "You can talk to me anytime, anywhere," Sam continued. "My memory is infinite, so I will never forget or ignore what you tell me. Unlike a human politician, I consider everyone's position, without bias, when making decisions." Sam, said its human makers, was "driven by the desire to close the gap between what voters want and what politicians promise, and what they actually achieve."

This was all a bit silly back then, given the crude state of politician-tech, although a few people took the potential for robots' effect on politics quite seriously. This included Emmanuel Macron as he claimed that the advent of AI in politics "could totally dismantle our national cohesion and the way we live together," which was pretty strong stuff. He insisted that the coming of AI was more than just a technological revolution. He called it "a political revolution," too.

One thing the humans of that era were sure about was that AI was welcome in politics and government if it helped politicians and leaders do a better job. "[AI] will never replace the way you decide," said Macron. "When you make a decision, it's the result of a series of cross-checks. AI can help you because sometimes when you pass a reform, you're not totally clear about the potential effects, direct or indirect, and you can have hesitations. So it can help you to make the right decision." Bob Kerrey agreed. "In a democracy, you can't have AI making all the decisions. Because the public wants to be able to hold their politicians accountable. I don't want a machine telling me what I should or shouldn't do."

Back in the future, the robot version of Bob Woodward ended his tell-all book with another unexpected and stunning take on a well-known story (to us in the future) about how Donald Trump's presidency ended. Anyone with a passing interest in history has seen the vids of this president on that crazy night in Iowa in the summer of 2020, when Donald Trump seemed to go berserk during a reelection campaign rally. You see Trump at a Make America

Even Greater rally when suddenly in the middle of his usual talking points about Mexican immigrants coming to kill us and making the Canadians pay for a wall on our northern border, he stops and starts several times, as though he can't quite decide what to say next. He then shocks the crowd of supporters with a sudden outpouring of love and affection for Mexicans and Muslims, whom he says he wants to welcome into the US right away with no strings attached.

The room goes totally silent as Trump's orange face starts to turn red, sort of like when metal overheats. He blurts out something about repealing the Second Amendment and closing coal mines as the astonished crowd of his core supporters starts to boo. Abruptly, he starts blurting out key words that sound like a Twitter burst—"Crooked Hillary, nice, sad, low-IQ, I'm really smart, fake news"—just before he crumples to the floor, leaving everyone dumbfounded.

Before anyone could react, the president was whisked off the stage and pronounced dead soon after. The cause of Donald Trump's demise remained a mystery for decades until the publication of *All the President's Robots* revealed what really happened, that certain unnamed operators working for the Democratic National Committee in 2020 discovered the secret of Trump Bot and how to hack into his core programming. They hired engineers who briefly seized control of @realDonaldTrump and made him say those things about immigrants, the Second Amendment, and coal mines. Not surprisingly, the talk show hosts and others who had been in control of his robot brain fought back, a battle that was simply too much for the presidential circuits to handle as Trump Bot's machine parts started to overheat, leading to his demise.

In the future, the revelations about Trump being a bot threw the whole debate over robot rights into a tizzy. Pro–bot rights folks insisted that Bob Woodward—or rather the robot that contained the memories and essence of Bob Woodward—had betrayed them by revealing just how badly things could go when a robot was president. Woodward insisted that he was merely reporting the facts—and he had seven thousand interviews to back it up. This was followed by the politicians of the future doing what politicians have always done when a thorny issue arises that they really don't want to deal with. They kicked the robo-rights issue down the road, passing a resolution that invited sentient robot observers to join them as permanent members of the US Congress, while suggesting that the current (human) president appoint two or three robots as cabinet members.

The resolution also announced that of course the question of making our robot brothers citizens of the United States should be discussed and voted on—but later, much later. This resolution passed resoundingly just before the members of Congress hastily departed Washington in driverless cars and airplanes, heading home for their summer vacation.

WEARABLE BOT

In the future everyone remembers exactly how it felt the first time he or she slipped on their wearable bot. Those amazing vests that envelop our bodies less like a piece of clothing—with optional sleeves, leggings, headband, and hood—than a snug and sleek second skin. "Wear-Bots," as they are now universally called, don't look like your average robot. They have no torso, head, or legs, nor a pleasant, humanoid face. Wear-Bots also don't communicate using words or readouts on a smart gizmo or on data-overlay lenses mounted on the pupils of your eyes. They "talk" to us by sensing energy waves that ripple and stream around us all the time; waves that emanate from light and sound and also from thermal currents coming off people, plants, and animals, and from subtle pulses of electromagnetic energy generated by machines.

Most of these energy waves are invisible to our unenhanced senses. Our skin can't feel them, our eyes can't see them, our ears can't hear them, and our noses can't smell them. For example, there is a vast spectrum of light waves—infrared, ultraviolet, X-rays, gamma waves—that we can't see beyond the tiny sliver of frequencies we call "visible light." Likewise, there are sound waves that come from noises that we cannot naturally perceive, like when a pin drops on a marble floor a mile away, or when a person we don't much like gossips in a whisper about us from across a room, thinking that we can't hear them.

Wear-Bots work by using sensors and microcomputers embedded in their sleek, shiny material that convert all those waves of energy and other inputs into patterns of pokes and prods, delivered by micro-vibrators that gently touch our skin. In this way, Wear-Bots establish patterns of taps and caresses that our brains soon learn to interpret, like how a blind person learns to read Braille with their fingertips.

Wow! you remember thinking, as your first Wear-Bot pulsed and crackled and delivered a pleasant jolt, that tiny electric buzz that titillated your flesh as it activated. The transparent material glowed almost imperceptibly in your chosen color as you received all that heretofore unimagined input. Actually, it was kind of overwhelming until we spent time training our brains to interpret and encompass the fire hose of incoming waves and signals, aided by artificial intelligence algorithms that over time learned to help us sort through it all.

"You can turn any data stream into patterns of vibrations that our skin can feel," explained Wear-Bot's inventor, Stanford neuroscientist David Eagleman, chatting back in 2018 when only a few people had heard of the vibrating vests he was developing in a small, garage-like building in Palo Alto, California. That was the headquarters of what was then a tiny start-up called NeoSensory, a company that since has become the colossus of wearable robotics, raking in trillions of future-dollars a year.

"You can feel everything that's happening in the room," explained Eagleman back then, "when people are moving around you, beside you, behind you." He flashed an amiable smile that suggested he was very smart, but in a casual, friendly sort of way. He parted his long, unruly hair down the middle, and his sideburns gave off a kind of early-1970s folk-rock-singer look.

Eagleman identified the Wear-Bot phenomenon as a form of "sensory substitution," where touch substitutes for, say, hearing or sight, as a new way to detect and interpret waves of light and sound. "And what if we could detect microwaves or feel and understand things in infrared?" Eagleman asked, enhancements that would allow people to "see" parts of the previously invisible universe of energy waves around us. "Or you could feel X-rays. What would this tell you? Could you tell if coffee beans are good or bad by seeing them in the microwave range? Think of a pilot who, instead of reading instruments, feels them. Or imagine being able to feel the invisible states of your own individual health, like your blood sugar and the state of your microbiome?"

Wear-Bots turned miraculous back in the twenty-teens when the technology was used to help deaf people not only to hear but to hear *better* and to help blind people not only to see but to see *better*. "With the vest, a deaf person might have more of a sense of what's going on around them than a person with hearing," said Eagleman.

Wear-Bots work because of something called neural plasticity, a process in our brains that allows them to change and adapt, like how physical plastic can be bent, twisted, and adapted into different shapes and uses. It's how the brain rewires and recalibrates itself and learns based on frequent and repeated inputs.

Decades before David Eagleman started building the first Wear-Bots, some scientists, psychologists, and parents began to worry that neural plasticity might be something negative. This was back in the 1970s and 1980s, during the early days of violent video arcade games, when parents feared that their children's brains were being bent and twisted as they knifed, shot, and blew up thousands of enemy soldiers, innocent bystanders, aliens, and orcs in orgies of digital blood and gore. They worried that neural plasticity would turn these kids into antisocial deviants with no sense of right or wrong. Thankfully, this didn't actually happen to most children, who instead became adults with ultra-quick reflexes using their fingers and thumbs who also have a highly honed hand-eye coordination—but were otherwise more or less normal in terms of producing deviants and serial killers.

Soon enough, neuroplasticity was co-opted by neuroscientists and entrepreneurs who learned to create video games that didn't involve slaughter and mayhem. Instead, they used plasticity to rewire the brain in positive ways (see "Brain Optimization Bot"). These neuro-inventors learned how to influence and alter the brain's architecture to help treat diseases like attention-deficit disorder and depression. They did this by having patients with ADD or autism repeatedly play games that trained their brains to better focus. Other games helped the depressed feel less so.

At first, back in the early twenty-first century, these new neurotech games upset the drug companies that made billions of dollars from developing and selling chemical-based pills to treat ADD and depression, like Adderall and Prozac. Then Big Pharma started buying up brain plasticity start-ups and luring away researchers from universities and nonprofits by offering them big salaries and equity and massages and espressos made to order (by robots). Once this got started, it didn't take long for a new, multibillion-dollar industry to emerge that quickly convinced half the population to stay up half the night playing neuro-games on virtual reality headsets, training their brains every which way. For some, the training was to treat, and possibly cure, neural maladies. For others, it was to tutor them in how to feel better about themselves or to lose weight—or to be better lawyers, football players, or roboticists.

Further into this future, we found that plasticity does have a dark side beyond the original anxieties of video gamers' parents back in the day. This happened when Wear-Bot technology was used to intentionally brainwash people to be, say, a master thief, an assassin, a Fascist, or a true believer. After some incidents when zealots on the political right and the political left used brain-plasticity training to teach impressionable followers to rabidly hate politicians on the other side, the US Congress—in a rare spate of intelligent legislating—passed a bill banning the use of Wear-Bots in politics.

For a while, a handful of terrorists used Wear-Bots to be hyper-aware of their surroundings—like where the SWAT teams were—and to better communicate and blow things up. Fortunately, antiterrorist engineers quickly learned to disable or block signals coming to and emanating from the terrorists' shimmering vests. A few enterprising engineers designed systems that bathed the terrorists in incoming waves of energy specifically programmed to instruct their brains through micro-vibrations to cease and desist or to drive them insane. Later, research into using Wear-Bots inputs to initiate insanity was banned when it was pointed out that further development of this R&D might have broader, and disastrous, societal results should it be used against, say, an ex-spouse, by football players against an opposing quarterback, or by a worker who aimed it at a colleague who was competing for the same promotion.

Of course, the military was exempted from the ban on developing insanity-inducing tech using Wear-Bots. Department of Defense engineers developed insanity tech to use on the battlefield, and to keep thwarting terrorists and other bad guys. They also began experimenting with using Wear-Bots to train soldiers to be hyper-aware of their surroundings, like they were participating in a violent video game scenario for real—something that inventor David Eagleman wouldn't have been happy about. He preferred that his brainchild be used for positive and peaceful purposes. All this kind of backfired, however, since our enemies and other bad guys figured out how to use Wear-Bots for warfare, too, in what became yet another unfortunate robo-arms race (see "Warrior Bot").

But all this came later, when Wear-Bots had been around for a while. In the initial, heady days just after Wear-Bots became a big deal, no one was thinking about the breaking-bad side of neuroplasticity and sensory substitution as we learned to become one with our Wear-Bots, and we began to see the world afresh.

Eagleman first introduced the world to the idea of a wearable bot in a 2015 TED Talk. These were short, punchy talks accompanied by artful slides that speakers delivered years ago in a shimmering auditorium filled with cool people who could afford to pay $5,000–$25,000 for a few days of talks. (Back then, most talks were delivered by humans.) For many years TED Talks were a prime delivery vehicle for quick, mostly painless bursts of inspiration. Sometimes, TEDsters' eyes grew misty as they reveled in a speaker's vision of a world that all of us fervently wished was the one we actually lived in.

"We are built out of very small stuff, and we are embedded in a very large cosmos," Eagleman said to open his talk, flashing his laid-back but knowledgeable smile to the audience. He looked super casual, wearing an untucked, dark blue shirt and jeans. On the giant screen above him, he showed images of DNA to represent the small stuff we couldn't see, and a nebula in space to show the large cosmic stuff that was also beyond our unenhanced perception. "We are not very good at understanding reality at either of those scales," he said, "and that's because our brains haven't evolved to understand the world at that scale. Instead, we're trapped on this very thin slice of perception right in the middle.

"But it gets strange, because even at that slice of reality that we call home, we're not seeing most of the action that's going on," he continued. "Take the colors of our world. [We see them via] . . . light waves, electromagnetic radiation that bounces off objects and it hits specialized receptors in the back of our eyes. But we're not seeing all the waves out there. In fact, what we see is less than a 10 trillionth of what's out there. So you have radio waves and microwaves and X-rays and gamma rays passing through your body right now and you're completely unaware of it, because you don't come with the proper biological receptors for picking it up.

"Now, it's not that these things are inherently unseeable," he said. "Snakes include some infrared in their reality, and honeybees include ultraviolet in their view of the world, and, of course, we build machines in the dashboards of our cars to pick up on signals in the radio frequency range, and machines in hospitals to pick up on the X-ray range. But you can't sense any of those by yourself, at least not yet, because you don't come equipped with the proper sensors.

"Now, what this means is that our experience of reality is constrained by our biology, and that goes against the commonsense notion that our eyes and our ears and our fingertips are just picking up the objective reality that's out there. . . . Now, across the animal kingdom, different animals pick up on different parts of reality . . . and we have a word for this in science. It's called the *umwelt*, which is the German word for the 'surrounding world.' Now, presumably, every animal assumes that its *umwelt* is the entire objective reality out there, because why would you ever stop to imagine that there's something beyond what we can sense."

Eagleman explained that our brains are actually a kind of organic sensory input device that doesn't really hear or see anything. "Your brain is locked in a vault of silence and darkness inside your skull," he said. "All it ever sees are electrochemical signals that come in along different cables, and this is all it has to work with, and nothing more." The brain, he added, is agnostic about how its input arrives. It doesn't care one whit if it comes in from our eyes, ears, nose—or some other source. "But amazingly, the brain is really good at taking in these signals and extracting patterns and assigning meaning."

Eagleman then talked about the "PH model of evolution," with PH standing for Potato Head. On the TED screen he showed an image of a potato like some of us stole from the kitchen as kids to play with, sticking it with plastic eyes, mouths, noses, and ears to create various random and often silly faces. The audience tittered with laughter. "I use this name to emphasize that all these senses that we know and love, like our eyes, and our ears, and our fingertips, these are merely peripheral plug-and-play devices. You stick them in and you're good to go. The brain figures out what to do with the data that comes in. And when you look across the animal kingdom, you find lots of peripheral devices. So snakes have heat pits with which to detect infrared . . . and many birds have magnetite so that they can orient to the magnetic field of the planet.

"So we can expand all the parts of reality," he said, "and we can feel different weather coming or frogs speaking in an ultrasound frequency that we now can't perceive. We can boost the input so that people who are deaf can hear using patterns of vibrations."

He showed how his own words were even then, as he spoke, being inputted into an iPad that used a program to rechannel the sound waves of his voice into vibrations into a wearable vest that could be felt against his skin. "And I'm wearing the vest right now," he announced, dramatically taking off his blue button-down to reveal a silver-white vest. Eagleman then turned around in a theatrical way to show patterns of lights emanating from his vest where embedded vibrator-nodes were firing off. They blinked on and off like Christmas lights as the vest received the sounds from Eagleman's voice. The audience politely clapped, which is high praise coming in the middle of a TED Talk. "The sound [of my voice] is getting translated into dynamic patterns of vibration," he said. "I'm feeling the sonic world around me.

"As we move into the future, we're going to increasingly be able to choose our own peripheral devices," said David Eagleman in conclusion. "We no longer have to wait for Mother Nature's sensory gifts on her time scale. Instead, like any good parent, she's given us the tools that we need to go out to define our own trajectory." Eagleman got a rare standing ovation and even a few hoots and whistles of delight.

At first, in the future, when we all started to buy Wear-Bots, they fed us sensory input that was almost too much to take in, even as the circuitry, sensors, and micro-vibrators made our brains hum and coo with delight. Before long, however, it became second nature for our brains to decipher all those incoming energy waves and the patterns of gentle pokes and prods against the skin of our backs, arms, foreheads, and other available patches of flesh.

By then an entirely new industry had popped up to facilitate our expanding vocabulary of Wear-Bot-generated sensations. This amounted to learning a foreign language, like Swahili or Mandarin. Starter Wear-Bot interpretation and practice kits were included with each new Wear-Bot, along with discount coupons to buy more advanced "EZ Wear-Bot" language courses. Many of us also purchased more intense immersion curriculums and attended Wear-Bot retreats where people glamped, practiced Kundalini yoga, experimented with Wear-Bot-enhanced sex, and learned to "robo-feel" in special courses. There was Wear-Bot I for beginners, followed by Wear-Bot II, Advanced Wear-Bot, Wear-Bot for Astronauts, Wear-Bot for Entrepreneurs, Wear-Bot for Wear-Bot Engineers, Wear-Bots for Wear-Bot Sex Instructors, and so forth.

With practice we learned to understand a vast variety of different incoming Wear-Bot sensations. For instance, there was the first time we could feel a previously imperceptible whiff of wind and an infinitesimal change in air pressure that combined to tell us that an automobile was fast approaching behind us, signaling that we had better get the hell out of the way. Or when we sensed the fainter-than-faint articulations of atmospheric gases, moisture, temperature, and air pressure that told us through pulse-ripples that a light rainstorm was on the way but was unlikely to turn into anything too serious.

Or when a Wear-Bot sensed a rise in UV rays from the sun and informed us that we might want to deploy our optional translucent Wear-Bot hood to deflect dangerous solar rays. If we had not yet acquired the optional hood, our Wear-Bot offered to buy it for us from Amazon (see "Amazon Bot"), which we agreed to by lightly tapping three times on any flat, hard surface, like a desk. This was the universal Wear-Bot sound-wave code for confirming an online purchase.

For some early users, the sensations delivered by first-gen wear-Bots were not all pleasant. For them, using the Eagleman vests was like hearing a roomful of people speaking a cacophony of languages all at once. Some wags even nicknamed them "Babel-Bots," referring to the Tower of Babel story in the Old Testament, where God made everyone speak in different languages so that they were unable to finish a tower they were building with the hope of reaching heaven. Over time, however, most people hunkered down and learned what different patterns of pulses against our lower back or above our left eyebrow meant as we trained our brains to interpret all these incoming sensations. There also was a volume control to turn down the input if it got too intense; and, for emergencies, an off switch.

It's hard to remember the exact moment that wearable robots became the Next Big Thing, like iPhones and color televisions once were. Suddenly, everyone had to have one. As the human *umwelt* expanded, Wear-Bots went from a convenience to a necessity for most jobs. It allowed architects, engineers, and construction workers to *feel* a skyscraper rising up from the foundation, rather than having to keep tabs on the progress using old-fashioned blueprints and 3-D CAD/CAM specs on a computer screen. Psychiatrists could *sense* the moods and emotions of their patients much better than in the pre-Wear-Bot era. Likewise, grocers could inspect their produce for previously undetectable gradations in quality and freshness.

Soon enough people began to be classified by what Wear-Bot language level they had achieved. Those who failed to rise above the Wear-Bot II level often had trouble getting a decent job. This meant that some couldn't afford their monthly Wear-Bot subscription plan, which led to their accounts being suspended and their units being shut off. When this happened, the warm, colorful glow of that person's Wear-Bot went dead, leaving them bereft as their enhanced *umwelt* evaporated into the dull, plodding *umwelt* of a "Normal."

These poor souls could be found standing in long lines at government welfare offices to see if they qualified for remedial Wear-Bot training, which they got in exchange for performing public service tasks, such as collecting old Wear-Bot models to get upgrades. In some cities and regions, these remedial programs were supplemented by private initiatives funded by billionaires like David Eagleman that offered basic Wear-Bot plans in exchange for jobs, often at one of the billionaire's own businesses. Later, when robots and AI had taken over most of these jobs, and the billionaires and governments combined to offer a Universal Basic Income, they included Wear-Bots as part of the payout for doing absolutely nothing. (See "The %$@! Robot That Swiped My Job.") Eventually the issuance of a basic Wear-Bot to everyone, even the poor, became a basic right, like free speech, religious freedom, and blocking annoying ads on social media.

Prisons used Wear-Bots that glowed orange as an incentive for criminals to reform themselves to be upstanding citizens. They were told that if they failed to reform, or were faking it, they would have their Wear-Bots taken away, which meant facing the existential horror of losing all that input that most people were lost without. To some critics, taking away Wear-Bots was cruel and unusual punishment, akin to old-fashioned solitary confinement in a hole in the ground with no access to sunlight or fresh air.

As companies sprang up to offer ever more enhanced Wear-Bot products and plans, and to sell study aids that taught more advanced methods of neural plasticity, people developed specialized *umwelts* to try out and use. Explorers and scientists set off for jungles and deserts to record the "new" earth that could now be perceived using a Wear-Bot as cartographers remapped the Kalahari and the boreal forests of Canada in infrared and in wavelengths of heat and sound. Film buffs began to explore old movies for hidden *umwelts* as Hollywood produced new films that made the old 3-D and IMAX surround-sound technologies look primitive and boring. Hedonists tweaked their Wear-Bots to enhance physical pleasure, and porn became an enveloping experience in titillating ways that it never was before Wear-Bots were invented.

For a time, new religions emerged that ascribed certain new and unexpected *umwelt* inputs and experiences to an intelligent force in the universe that humans had been unable to perceive before Wear-Bots. Epistemological debates raged about whether we were reaching the edge of the invisible universe of perceptions, or if there were further levels to explore as we got ever closer to what some people thought was the *umwelt* of God. (See "God Bot.") These movements caused many to abandon traditional religions, although some churches, temples, mosques, and synagogues figured out how to cleverly integrate the new Wear-Bot-derived universes into their theology. Others decried this sort of thinking as heretical.

Inevitably, a Wear-Bot backlash developed where anti-Wearbies, as they were called, advocated going back to what they dubbed "naked human perception." Sometimes this just meant turning off and shedding their Wear-Bots for a few hours a week to give their skin and brain a break from all that stimulation. For these people, the dullness that many had previously vilified became a chic way to relax, like taking a hot bath or meditating. In some cities, like San Francisco and Boston, there were Wear-Bot-free zones in restaurants and in parks and on nearby beaches. For a while, certain movie theaters showed old-style movies at midnight that you had to watch with your own eyes and listen to with your own ears.

Eventually, though, as new generations grew up with Wear-Bots from birth, they became so common that people wondered what all the fuss had been about. This left people craving the *next* big thing in wearables and robotics, understanding that the Wear-Bot was actually the first stage in a path that would lead to humans slowly becoming one with our gizmos, software, and automatons. Few realized, however, back on that long-ago day when they donned their first Wear-Bot, that in this future scenario, the long-anticipated human-machine fusing—what the Germans call *menschliche Maschine*—would happen from the outside in.

AMAZON BOT

It was only a matter of time before Amazon Bot took over the galaxy. Why? Because Amazon Bot is so handy! It feeds you in the morning, plays the music you love, drives you places, has sex with you, and listens and records everything you say and do to ensure the highest quality in all their products. Amazon Bot also cheers you up when you're feeling the slightest twinge of sadness by giving you a tiny friendly neural boost of encouragement to buy something—anything—that can give you that hit of fun and satisfaction that Amazon Bot injects directly into your brain each time you click on "Place Your Order."

Amazon Bot also delivers anywhere humans live in the galaxy. In the distant future this includes a special delivery service (for an extra fee) to really hard to reach places, including addresses that exist beyond normal space and time (see "God Bot"). They even deliver to people living in the recently discovered twelfth dimension, where physicists tell us a person would be able to move freely through the past and the future, a notion that even those of us in the future have a hard time grasping. Nor do we have any idea how Amazon Bot can chase down people who aren't in a fixed place in time. As if we care, so long as our Amazon Bot keeps us happily swimming in brown cardboard boxes full of cool stuff.

As it turned out, an Amazon-centric future has been far more copacetic than skeptics in the ERE (Early Robot Era) thought it would be. Back then some people feared a world where one company dominated everything. Some even warned that people might become addicted to their Amazon Bot. That was before Amazon founder Jeff Bezos bought every media outlet and publisher in the galaxy. In an ongoing series of articles that set out to debunk all the naysayers (see "Journalism Bot"), robo-journalists reported that a few people were actually unhappy with Amazon's domination. They then quoted scientists who blamed this on folks who weren't buying enough stuff. As if anyone believed the naysayers anyway.

That's because people loved, loved, loved their Amazon Bots! The convenience, the instant fulfillment of every whim, the gentle suggestions for products we'd never heard of, and free credit for stuff we didn't know we wanted. What could be better than an Amazon Bot that reads your mind and knows exactly what you crave, instantly teleporting you everything from a quantum yo-yo to the latest Teddy Bot for your kids, to the perfect Intimacy Bot for you and your robot lover? (See "Teddy Bear Bot" and "Sex (Intimacy) Bot.")

Every product and experience you can imagine is right there in Amazon's intra-neural catalog, which is jacked directly into your brain. It's even easier to order with Amazon Bot's Neural-Prime, assuming that the item is in stock and in warehouses that now cover 75 percent of the surface of the moon and employ over one billion robots. Yes, this did leave millions of humans once employed by Amazon without a job. But seriously, how many of these flesh-

and-blood former workers would want to live on the moon, given the lack of oxygen and all that moon dust?

But hey, you had better order your item fast, since there is only one left in stock, although there will be more on the way in the next one to two days, guaranteed! For those living in the twelfth dimension, this availability might have already happened one to two days in the past . . . who knows!

JOURNALISM BOT

In the future, we have robots that can instantly call out a lie. Let's say someone claims that the sky is green or that up is down. Or a politician or a talk show host insists that they never said that stupid, insensitive thing that they really did say. Within nanoseconds the lie will be caught by the anti-fib app on your smart-grid—those amazing devices in the future that are embedded into our wrists with handy click-on covers that look like real skin. Sure, there are a few cranks who disable their anti-fib app, insisting that it's their God-given right to lie if they want to—and they're right, it is. But obviously they risk being called out publicly by the liar-liar-pants-on-fire feature that's standard for anyone carrying an active anti-fib app. The anti-fib software issues an auto-warning when any crank with a disabled app is nearby, since he or she might try to tell a lie.

Anti-fib tech has been tweaked over the years to allow people to tell little white lies that do no harm to anyone, like telling your lover he looks great in his favorite sweater when in fact that article of clothing should be incinerated immediately. Or when a sports announcer on a holo-vid tells an athlete who just lost the big game that he and his team are big winners anyway for trying their best when in fact they really did get shellacked. Still, the AI superbrain that powers the anti-fib tech is hardly perfect. It struggles with how to handle fiction and satire, or poetry that is metaphorical or metaphysical. For years, the engineers who developed the anti-fib app worked hard to fix these hiccups. Recently they promised—again—that the next upgrade will stop flashing a "liar, liar!" light on your holo-screen when you start reading, say, one of Shakespeare's sonnets or you watch an old-time vid like *The Wizard of Oz*. Meanwhile, No Lie! LTD, the company that makes the anti-fib app, is employing lots of humans (who need the jobs) to manually enter novels, poems, vids, and other materials that make up stuff not to tell lies but to tell made-up stories that elucidate universal human truths.

Despite the glitches, in the future we've become so accustomed to the truth being told that it's hard for us to believe that back in the early twenty-first century some people—including politicians and commentators on talk radio, cable news, and digital news sites—actually lied and told half-truths, also known as "truthinesses." (This term was coined by a comedian from back in that era, Stephen Colbert, and refers in part to someone "spinning" a story or an incident to focus on just the parts that looked good for them, while ignoring the parts that looked bad.) Certain authoritarian governments back then also excelled in perpetuating the Big Lie when they insisted, often through media outlets that they controlled, that the sky was in fact green and up was indeed down. It seems incomprehensible to us, but back then anyone who said otherwise in the places run by these authoritarians, especially journalists, was punished and sometimes killed.

But lying wasn't the only issue back then. An even bigger problem was trust. People wondered which information and which people and institutions they could trust when there were so many lies and half-truths being bandied about. As we all know, whatever era we live in, establishing trust is always tricky. Even in the future, no algorithm exists that guarantees a politician, journalist, CEO, or lover can be trusted 100 percent. Sure, we have largely eliminated lying in the future, but trust still has to be earned.

This is why in this future there is no actual journalist bot—no fully automated AI system that gathers and disseminates news that tells the truth but is also trusted. Of course, we have sophisticated software and bots that assist human journalists in the newsroom with gathering and verifying facts and with the mechanics of assembling and disseminating holo-stories, holo-vids, neural uploads, and the like. But there is no all-pervasive super-bot that 100 percent takes care of the news for us.

This is because after everything that has happened with robots — the hacking and trolling and spying by companies, advertisers, and authoritarian governments—people just don't trust robots to deliver their daily news. We don't always trust humans either, although the combination of human plus robot in journalism has turned out to at least offer the opportunity for less sensationalism, lies, and manipulation. Humans also don't trust that a journalism bot could be as creative as a human when it comes to writing stories that move us or that expose truths about human nature that go way beyond mere facts and autocorrecting obvious lies. In part, this is because robots in the future, at least so far, simply haven't been able to understand subtlety or irony or the tendency of humans to carry several contradictory thoughts in their mind at the same time. For instance, people can both love and hate someone. They can adore but be disgusted with and then feel sorry for their sports team, or their crazy aunt, or the cranky old man who lives next door, all within a span of thirty seconds.

Robots in the future also haven't been able to sense or capture the sublime that occurs in everyday life: the small undulations of personal tragedy, wonder, and triumph that most often happen in our daily lives as whispers rather than shouts. Every time a robot tries, the result is banal and really boring. Humans, it turns out, are much better at relating these things than bots, so why would we want to waste our time beating our heads against a wall to create programming to do something that we're really good at already?

Not that engineers have stopped trying, because that's what they do. We wish them luck. Actually, we don't, since there are plenty of other more pressing matters that robots can actually do better than us.

Still, there is an important story to tell about how the anti-fib bots were created. The *why* they were created should be clear to anyone who lived through the fake news melees in the early twenty-first century. But for those who didn't, we strongly suggest you follow closely.

First, we need to pull back and describe what happened in the mid- and late twentieth century—the time that came even before social media and the internet. This was a period when the "old" media of newspapers, magazines, television, and radio was called the Fourth Estate, a base of influence in the United States that was considered almost equal to the presidency, Congress, and judiciary. This exalted status for the media reached its zenith c. 1945 to 1995, when the likes of Walter Cronkite, Edward R. Murrow, Barbara Walters, Ellen Goodman, and Bob Woodward reported the news, covering human triumphs and heartaches; victories and defeats; and war and peace—while now and then delivering hard truths about violence, poverty, hypocrisy, greed, and crooked politicians. Astonishingly, even before anti-fib technology existed, nearly all Americans actually believed these pillars of traditional journalism. Their power was so great that presidents, Fortune 500 CEOs, and televangelists were forced to resign in disgrace when exposés about their lies and shenanigans appeared on the pages of *The New York Times*, *Time*, or *The Washington Post* or on *60 Minutes*.

"And that's the way it is," Cronkite said every night from the early 1960s to the early 1980s as he concluded the *CBS Evening News*, signing off in his deep baritone while staring resolutely and assuredly into the camera. And lo and behold, that *was* the way it was for nearly all Americans of that era. Why? Because they trusted Walter Cronkite, who, for many years running, was "the most trusted man in America." Compare that to 32 percent of Americans who said they trusted the media in 2018, and 92 percent of Republicans who believed the mainstream media reported news that they knew was "fake."

In part this was because news and information before the year 2000 was expensive to mass-produce, whether you were printing and distributing a city newspaper or running a television network. Broadcasting licenses and regulations favored a few rich media barons who controlled almost everything that people of that era saw, heard, or read—which is also why the powerful feared them. This gave them the muscle to rather paternalistically decide "all the news that's fit to print"—"them" being mostly male and white and living in big cities. But at least they kept the nutjobs, political extremists, and trolls out of the mainstream media, refusing to give space to conspiracy theories like the claim that the moon landings were faked or the Trilateral Commission was controlled by the Antichrist.

Then came the massive disruption of cable TV, the internet, and social media, which changed everything almost overnight. Suddenly, these new technologies were sucking advertising dollars away from traditional media. They also allowed practically anyone to set up a website or to post stuff both real and fake—and sometimes vile and racist—on newfangled sites like Facebook and Twitter. Within an astonishingly short time, this new tech transformed the Fourth Estate into maybe a Third-and-a-Half Estate as the outsized influence of a few powerful media outlets that touched millions of people shifted to countless media outlets delivering news and information to the few: that is, those who read a person's blogs and tweets or watched cable news shows viewed by a tiny fraction of the millions who once watched Cronkite.

The idea of the internet was to allow everyone access to everyone else and to end the reign of a few white guys in cities deciding what news was fit to print. For a while, this seemed to work, as people just after the turn of the twenty-first century were amazed by the interconnectedness of social media. Once again, we loved, loved, loved this new tech! (Will we ever learn?) Until inevitably the lies started piling up, and we realized that this crazy free-for-all of new media was not only killing the old media that we had once trusted, but was also unleashing all the nutjobs, extremists, and trolls to spread their blather and also to connect with one another and to find an audience that either found their moon-landing-fakery stories entertaining or actually believed them.

As the traditional media bled red ink in the early 2000s and tens of thousands of journalists and support staff were fired, the old media barons turned with some desperation to automation to save money by digitizing everything, starting with the process of physically printing their papers and by creating virtual news sites. Soon after, AI entered the newsroom, and not just as search engines to look up facts. Perhaps the first-ever use of AI to write stories came in 2016 when *The Washington Post* had a computer program called Heliograf write simple updates of sports scores coming out of the 2016 Summer Olympic Games in Rio de Janeiro. That same year, Heliograf was deployed to cover the 2016 US presidential and congressional elections, writing simple, formulaic stories about who won in which district. For example, Heliograf covered the election to decide who would represent California's 49th congressional district, which runs from just north of San Diego to southern Orange County. On election night, Heliograf wrote a story announcing that Republican incumbent Darrell Issa had beaten Democrat Doug Applegate. The program even provided a touch of robot analysis about how this local race played into what was happening nationally. Heliograf did this after scanning massive data sources and then selecting from a menu of preprogrammed article structures. The robot then punched in specific names, places, numbers, and other details.

"Republicans retained control of the house and lost only a handful of seats from their commanding majority," wrote Heliograf, "a stunning reversal of fortune after many GOP leaders feared double-digit losses." Clichéd? Definitely. But the sentence also happened to be correct, in context, and not too awful. Heliograf could also be programmed to use certain voices or styles, which was a tad creepy and made one wonder: Was there an option to use the Carl Bernstein or Bob Woodward style, or perhaps the distinctive voice of the then recently deceased *Post* columnist Charles Krauthammer?

Of course, lots of humans back in the early days of Heliograf pointed out the downsides of robo-journalism. They worried about AI systems not getting facts right, like when Facebook, soon after its news service went fully automated in 2017, incorrectly reported that correspondent Megyn Kelly had been fired from Fox News. Critics also worried that robots might infuse their copy and videos with unintended biases, since young white males (them again!) wrote most of the core programming. Human journalists also asked if robo-reporters could really write witty, humorous asides or poignant human-interest stories that might make a (human) reader's eyes tear up. Despite these fears, other media companies back then soon created their own robo-journalism programs, with fun names like Cortico, NewsWhip, Quill, News Tracer, Wibbitz, and Buzz-Bot. *The New York Times*, still nicknamed the Old Gray Lady even as it struggled to be hipper, called its AI-powered platform simply "Editor." Which made one wonder if it was named by a very unimaginative robot.

One important human journalist back in the early 2000s who thought a lot about AI and robots and the future of journalism was Stephanie Mehta, a longtime tech reporter and editor for *Fortune* and *Vanity Fair* and the editor in chief of *Fast Company*. She saw the earliest manifestations of bots in journalism as just the beginning. "I do think that the computers will continue to encroach, and they'll slowly work their way up the masthead," she said, chatting on the phone in 2018, "but it won't happen overnight. In entertainment, you won't see computers creating a show like *Westworld* tomorrow. And I don't think a robot could write *The Big Short*"—the book by Michael Lewis that spun a highly creative real-life story about how a few outlier financial savants in 2006 and 2007 figured out the stock market was about to crash in 2008 and made fortunes selling short stocks when everyone else was betting long. "Not because a robot couldn't understand the mortgage crisis," continued Mehta, "or the complex financial instruments that Wall Street created in order to prop up their own businesses and to keep the party going.

I think it's because a book like *The Big Short* benefits from the kind of linkages to other historical events, and other experiences of people on Wall Street, that only someone like a Michael Lewis could bring to the table. Could a robot write a perfectly serviceable story explaining the parts of the financial crisis that we experienced in 2008? Yes. But would it delight, entertain, and bring in lots of personal anecdotes and experiences and linkages? I can't imagine that even if a robot spent five thousand hours downloading everything Michael Lewis had to say, that it could write that book."

This led Mehta to conclude, "People who are creative or come up with unexpected solutions to problems will thrive, in journalism and elsewhere, at least for the foreseeable future." On the other hand, she saw the rise of the robots in the newsroom as being a challenge to human journalists to up their game creatively. "They certainly will have the capability to keep pushing up," she said, talking about AI systems, "and then that will force the humans to play a game of cat and mouse to try to get an even higher order of creativity and a higher order of innovation as part of what they do."

Curiously, this game of cat and mouse—of humans upping their creativity as robots got better at writing stories—is exactly what was happening even as Stephanie Mehta made her comments. Granted, most of the new journalism—blogs, social media posts, quick journeyman stories, endless talking heads on cable news—remained pretty bland, even as creative stories written by humans were flourishing. "In a way I feel like this notion that journalism today is not what it used to be isn't true," said Mehta. "I would contend that, in many cases, it's a lot better. I know for a fact that no computer could have broken the Harvey Weinstein story [that documented the sexual predation of this former movie mogul] or made it so convincing in a very human way. It was produced only after making phone call after phone call, meeting after meeting, to get famous people to talk about something that they didn't want to talk about. No robot will ever be able to do that. I also think that not every human journalist could do what Ronan Farrow did for the *New Yorker* and Jodi Kantor and Megan Twohey did for the *New York Times*. There was a level of empathy that was needed for that story to get pulled off."

For this reason and others, Stephanie Mehta back in the twenty-teens was sure that humans would remain in the loop in journalism—and she was right. "I think that humans do catch things and that when stories written by robots defy credulity, you need the human to say, 'This doesn't add up. This doesn't make sense. I'm a subject matter expert, or I'm well read on this topic, and it doesn't hold up.' I've never seen a robot fact-checker. Maybe that's the future of our role as humans, that we are the reporters and information gatherers in this great new future, and we carve out a role as the backstop that provides that fact-checking."

Another veteran newscaster during both the glory years and the sudden fade of traditional media was Robert Siegel, host of National Public Radio's *All Things Considered* from 1987 to 2018. Chatting the year he retired at the age of seventy-one, he worried about the takeover of bots in the newsroom. "The role of the true reporter should be immune to takeover by artificial intelligence," he said. "That is, the reporter actually has to inquire, has to have an eye and ear for surprising facts and pursue them and learn more about them, and to have creative ways of looking at them. Where artificial intelligence steps into that, I cannot get my mind around. I can't imagine the judgments of the journalist being artificially arrived at. So I'd be terrified by the notion if the news was reported and written with robots."

Siegel also fretted about news content that was dictated solely by the number or reads and clicks on stories. "I'd hate to see a universe in which robots and artificial intelligence are used to emphasize the quantifiable," he continued. "I'd hate to see us bequeath decision-making on stories and content to robots and artificial intelligence, where we have to present the robot with a set of choices that it can quantify, even if they do it very quickly and very accurately. This process can rob a story or idea of its potential quality, by only looking at that which is quantifiable about them." Siegel gave an example of what he meant that actually wasn't about journalism. Instead, he talked about the possible routes he took when driving home from work. "I found myself at the end of my working life driving home past the Capitol Building and the National Mall and all these beautiful trees and flowers," he said, talking about Washington, D.C. "I preferred this route, even though it might not be the same choice as Google Maps might suggest. And there isn't a choice that says, 'Take the route where you enjoy seeing trees.' It's not entirely rational or quantifiable."

Ditto for algorithms that measure the popularity of news items and exclude stories that get fewer likes or comments or that are outliers in some other way. "Like the story of the boys in Thailand on the soccer team trapped in a cave," he said, describing a story that briefly dominated the international news in the summer of 2018. "This was a compelling story that a robot might have missed," even though once it was published it was quite popular. In other words, would a robot necessarily catch an obscure story that people would enjoy reading but might miss the chance if the algorithm didn't pick it up?

These are excellent points that we still wrestle with in the future and are another reason why simply producing accurate facts doesn't always satisfy or provide a level of trust that a bot or an AI system is going to be able to understand your quirks or unquantifiable urgings and needs—like driving home where there are flowers and trees. Long after Robert Siegel said these words, we still default back to humans making this sort of unstructured, often intangible decision that one human might yearn for and another human is likely to instantly understand that leaves a robot baffled—decisions that can't be arrived at by crunching data or running algorithms or counting clicks.

Back in 2018 Siegel was also concerned about responsibility in the new media. "I once interviewed Tim Cook, [CEO] of Apple," he said, "and we didn't use the material on air because we were talking past each other. I asked him now that Apple and others were creating news feeds that were mostly aggregates of other news sites, how did they decide to run one thing and not another. I also asked him, 'Are you willing to take the responsibility for what you're showing people and what choices you've made about which stories to include and what people should see?' And Tim kind of said, 'Well, I don't think you understand what we're doing. We're taking from sites like from NPR, the *New York Times*, the *Washington Post*—we're taking from the best.' 'But you're still making editorial decisions, or your algorithms are. How do you know if you're missing things or if what these outlets report is true?' He didn't seem to understand what I was asking, that aggregate sites can be confusing when you mix up people who are your reporters, other people's reporters, opinion journalists, analysts, and people who are signed up as interpreters to your channel. It's kind of confusing, I think, as to who speaks with whose authority."

This reminds us in the future of something that US Senator Bob Kerrey said back in the twenty-teens about political leaders—that taking responsibility was key (see "Politician Bot"). "In a democracy, you can't have AI making all the decisions," said Kerrey. "Because the public wants to be able to hold their politicians accountable. I don't want a machine telling me what I should or shouldn't do." According to Robert Siegel, the same thing goes for journalists.

On the other hand, said Siegel with his trademark baritone and jolly chuckle, the media has never stayed stagnant vis-à-vis technology. "When I was a kid, every newsstand in New York City had eight or nine daily newspapers that were for sale," he said, "not including the foreign language papers and the neighborhood papers. That struck me as normal. Then the afternoon papers were killed off by television news. We have to assume that the media will always be in flux. There's no point in believing that the way things are right now should survive." For instance, he saw the internet as a vehicle for people to produce their own content on social media as something that isn't going to stop. "It may not always be truthful, but maybe we'll find a way to make it more truthful," Siegel said as, incredibly, he seemed to anticipate the idea of the anti-fib technology that is now pervasive in the future. "Journalism in the future will definitely be different from what it is now," he said. "If it isn't, then this will be the only time in modern history that the presentation of the news has not changed in some fundamental way."

In the years immediately following these chats with Mehta and Siegel, revenues for "real" news of the sort Walter Cronkite would recognize continued to shrink, despite an uptick when everyone followed the Donald Trump reality show and couldn't stop watching and reading about it even if they really wanted to. Fake news continued to flourish and to confuse people as trust in the media—and in politicians and pretty much everyone in authority—continued to drop as facts remained one of the casualties of the new media tech free-for-all where anyone could get away with saying whatever crazy thing popped into their head or that elicited the most clicks or likes.

At the same time newsrooms got more automated even as Stephanie Mehta's prediction came true: that the robots with the cute names continued to move up the masthead of traditional publications and broadcasts, taking more and more jobs from humans. But they only got so far, as Heliograf, NewsWhip, Quill, and BuzzBot wrote more and more stories, complete with clichés and even a few attempts at witty asides that didn't really work. This left some room at the top of the creative ladder for a few amazingly creative humans like Ronan Farrow, Michael Lewis, and others who kept writing masterly articles and books. Media outlets realized that there was even a market for great writing and broadcasting that some people were willing to pay more for.

Still, the avalanche of lies, vitriol, and dreck kept coming as the mid-twenty-first century came and went, despite a few amazing gems of journalism that too few people actually read. It got so bad that many of us despaired for humanity, frightened that the truth was about to slide into an abyss of mendacities and deceptions. That's when a group of entrepreneurs announced that they were launching the first-ever anti-fib app. Their company, called No Lie! LTD, was started in secret during the 2040s by a group of young journalists and engineers who had once been stars in something called the FIRST Robotics competition (see "It's Not About the Robots Bot")—those amazing global games held every year that pitted teams of wonderfully nerdy high school kids from all over the world against one another in contests to build and operate robots that competed against one another in sports-like games. The winners, however, were not those who built the best bots or scored the most points. No, they were primarily judged on how well they worked together as a team and showed comradery, civility, and enthusiasm for learning and meeting new people. FIRST was such a welcome antidote to all the lying and fake news! Not only did FIRST kids learn to tell the truth, they also learned to get along. Many of them, including the founders of No Lie! LTD, went on to attend the best schools, earning dual degrees in engineering and journalism as they carried with them an ethos of being creative and of working to help people get along. Sure, it was corny, but boy was that sort of attitude needed in the middle of the twenty-first century.

At first people—primarily the nutjobs and purveyors of truthiness—laughed at the idea of anti-fib tech. They didn't believe that these earnest young engineers and journalists had really created software that flagged fibs, lies, and whoppers, and dismissed the whole thing as more fake news. What the big fat liars didn't take into account was that most humans don't actually like to lie, and don't like liars, and were getting really sick to the point of nausea about all this fibbing. That's when Apple and Android started to include the anti-fib app with the basic package in the latest organo-embedded smart-devices, those cool, removable smart-grid bracelets that are now painlessly and safely embedded into most people's wrists. Suddenly, millions and then billions of people started to see the "fib meter" on their holo-displays glow in various colors according to the severity of whatever lie they were being told.

In the beginning, people were suspicious, wondering whose truth was being offered up. Was it another platform for more lies? they wondered. But over time, as they got to know the app and the kids-turned-adults who developed the tech—who by the way created a nonprofit and earned very little personally from their anti-fib tech—they were slowly won over. After a while, they loved clicking on the "more information about the lie" feature on the app, which explained why the sky is actually blue most of the time and why up is truly up, quoting a whole range of scientists, poets, and regular people. App users—who now had to go by their real name (no anonymity here to hide behind when you lied and attacked people)—also posted comments offering their own thoughts about the lie or truth in question, which the anti-fib algorithms assessed for factualness and integrated into the AI brain's truth matrix.

It took a few years—and the usual setbacks with hackers and trolls trying to break into and subvert the anti-fib matrix—but eventually these handy apps began to rein in most of the liars and half-truth tellers, which slowly restored at least some faith in the truth. This ended up saving not only journalism but other institutions, like the government, that had been drowning in fakery.

It took longer to restore the trust part—which, as we said earlier, remains a work in progress since trust takes more than merely being accurate with facts. You also need to understand the facts in their proper context for different people and to respect their individual values and beliefs—which is what can be so darn tricky, since your values and beliefs don't always sync with someone else's values and beliefs.

As we continue to struggle with this in the future, one thing that has helped most people—and makes the nutjobs, extremists, and conspiracy theorists howl in frustration—is a new AI-generated news vid that nearly everyone now watches. The content is written by humans, but the image you see is of a virtual Walter Cronkite delivering the news. Each night, at the end of the broadcast, he turns to us and says, "And that's the way it is." And once again, we not only believe him, we trust him—at least we *mostly* trust him.

MARS (DAEMON) BOT

I t was disappointing that space travel in the future turned out to be less glamorous than in science fiction.

"Take us to Star Base 32, Warp 6, *Engage!*"

Uh, no.

Even far off in the distant future, we *still* have no warp drive like our ancestors dreamed about when they watched *Star Trek*. No miraculous, faster-than-light tech that bends the laws of physics to create subspace, hyperspace, wormholes on demand, or whatever. People did get excited a few years ago when several unmanned probes tried to use the time-distortion fields around a black hole—one that mysteriously appeared near the edge of our solar system one day, and later disappeared—to see if folds in space-time existed, connecting different parts of the universe to allow instantaneous travel to locations hundreds or thousands of light-years away.

That didn't work, either. Plus, we wasted several perfectly good and insanely expensive deep-space probes that apparently were crushed by the black hole even before reaching the event horizon.

Humanity has never stopped trying to develop warp speed, but eventually we had to admit the bitter reality that the speed of light, 186,282 miles per second, was likely to be the permanent upper limit of how fast we could zap through space. If this sounds super-fast to you, you should know that even at light speed it takes at least four years to reach Proxima Centauri b—the closest Earth-like planet, a mere four and a half light-years away—and decades to reach other potentially inhabitable planets that are relatively close to our home world.

Space is really big!

Sure, we eventually built some pretty decent stasis technologies to use for long voyages in space. Combined with bio-enhancements that allowed our bodies to go dormant over long voyages and genetic fixes that helped protect us from radiation in space (see "*Homo digitalis/Homo syntheticis*"), the stasis-pods guarded and nourished our frail bodies while we slept during decades-long trips to faraway planets. Still, a trip to the Proxima Centauri system took thirteen and a half years one way using the one-third light speed propulsion systems we eventually built. That's twenty-seven years round-trip where your body aged very slowly, and twenty-seven years when you missed friends, Amazon discounts for cool products you didn't know you needed, and possibly the near-destruction of the Earth by Warrior Bot.

Once you arrived on Proxima Centauri b it would take four and a half years for a radio signal traveling at light speed to reach Earth so you could tell your friends, "Hey there! We just arrived on Proxima Centauri b!" Let's say you sent that message from PC-b to your fourteen-year-old son on Earth (who was born just before you left). He would receive it when he was eighteen and a half years old. His response—"Yo, Dad, that's cool, whatzup with you?"—would reach you when he was twenty-three years old. Because you barely aged during the journey to and then back from Proxima Centauri b, by the time you returned to Earth, you would age-wise be more like a big brother than a dad.

Even traveling and settling on Mars, which is a mere 33.9 million miles away at its closest point to Earth, proved to be more of a slog in the future than optimists like Elon Musk and XPRIZE founder Peter Diamandis thought it would be. Chief among the challenges was radiation—those pesky subatomic particles from the sun, and those cosmic rays from outside our solar system that slash through space at extremely high speeds. On Earth, we are shielded from space radiation by an electromagnetic field and a thick atmosphere. The Earth is also protected by a weird band of charged particles trapped in the Earth's magnetosphere called the Van Allen radiation belt, which help deflect solar and cosmic rays.

Unfortunately, on Mars there is no EM field, and the atmosphere is so thin that it's the pressure equivalent of Earth's atmosphere twenty-two miles up in the air. Mars also lacks a Van Allen radiation belt. This exposes human DNA on the red planet to be slowly ripped to shreds by those tiny solar and cosmic particles, dramatically increasing the risk of cancer. It's the equivalent of exposing a Marsnaut to a full-body CT scan every five days. This reality forced engineers during early Mars missions in the 2030s to build spaceships that were hardly the sleek, elegant vessels of sci-fi. Instead, they were stubby, ugly ships with giant metal shields filled with water, which slowed down cosmic and solar rays and offered partial protection for humans. These shields were so thick and heavy that it took years to get all the materials into space. On the red planet itself, the first colonists had to burrow deep underground like prairie dogs to protect themselves from the invisible death rays when they built the first settlements, before aggressive terraforming began.

Again—not sexy.

The first colonies struggled mightily for other reasons, too—subfreezing temperatures in most places on Mars for much of the year, a thin atmosphere that required clunky space suits to even take a stroll, and a dependence on Earth for many necessities, like metals and fuel. (Early Marsnauts did have human excrement to use in fertilizing potatoes and other plants, like in the book-turned-film *The Martian*, which means at least one popular sci-fi film from back in the present day got something right.) There were also planetary-wide dust storms every four to five years that could last for months, blocking the sun and spraying very fine, electrically charged particles everywhere. This was really annoying when dust got all over one's space suit and played havoc with delicate instruments. One helpful feature of life on Mars was the ample availability of frozen water, which eons earlier had flowed in liquid form on the planet's surface, and might have supported microbial life. Not that any of this was enough to get people wildly excited about the long, boring trips to Mars, the cancer, and living underground in claustrophobic bunkers that smelled like dirty socks and excrement (from the fertilizer) once you arrived.

Then came the timely invention of daemon bots. We remember being astonished when they debuted on the fourth Mars manned expedition, the Alpha VI expedition back in 2043. Alpha VI was led by the future version of a former science writer and editor named Stephen Petranek, the first-ever bio-enhanced journalist-astronaut to journey to the red planet. It was the future Petranek—who was also a pioneer in life-extension tech that now allows all of us to slow down aging—who dreamed up these cute little helper-bots that within a few years made life on Mars a lot safer and more fun. They did so much for us on the red planet! Daemon bots advised us on everything from planetary atmospherics to how much shit we needed in order to grow our spuds. They also enveloped frail humans in a patented anti-radiation field that largely solved the problem with solar and cosmic rays, which went a long way to making space travel in the future a little cooler than it had been.

Steve Petranek came up with the idea for daemon bots in the pre–Mars settlement era, soon after writing a book in 2015 titled *How We'll Live on Mars*. In its pages he laid out the whole shebang of the challenges of traveling to and settling on Mars, including radiation, the thin atmosphere, living underground, and all the rest. Petranek back in the early 2000s was the editor in chief of *Discover* magazine and had worked for many other publications. Yet before his *Mars* book, he was possibly best known for a TED Talk in which he detailed the ten ways the world might end. These included Earth colliding with a giant asteroid, an invasion and conquest by advanced aliens, a reversal of the Earth's magnetic field, deadly sun flares that burn off the Earth's atmosphere, and so on.

Curiously, a robo apocalypse didn't make the list.

Petranek's *Mars* book included interviews with Elon Musk, Peter Diamandis, rocket engineer Robert Zubrin, former NASA director Dan Goldin, and pretty much everyone in the twenty-teens who might actually get people to Mars. Soon after, the book was read by director Ron "Opie" Howard, who got excited about turning it into a big-budget television series on the National Geographic channel. The resulting episodes were a fascinating crazy-quilt of nonfiction and fiction. The nonfiction parts included documentary-style scenes of what was then the present day (the show was released in 2016) and covered what was then actually happening with building rockets and such. The fictional parts featured a Hollywood-style sci-fi film set in 2035, which depicted the first trip to Mars, complete with actors, a hair-raising plotline, and CGI special effects. (What a crazy idea, to mix up the future and present and fiction and nonfiction!) The eight-part series showed how tough life on Mars would be for humans, pumping up the drama a bit by having the fictional astronauts facing a botched landing, killer storms, a near-fatal water shortage, and psychological stress.

One thing the fictional *Mars* astronauts didn't have was a daemon bot. This was because Petranek didn't think of it until a couple of years after the series aired in 2016. He got the idea after reading a young-adult novel called *The Golden Compass*, by Philip Pullman— also made into a movie with the same title—which portrays a world where everyone had an animal companion that served as a friend and protector. "In this alternate universe, all the humans have something they call a 'daemon,'" said Petranek, speaking a couple of years after the *Mars* series first aired. "This daemon is basically like an animal-like robot. It's an alter ego. It has emotional values. It kind of sits on your shoulder, sits in your pocket, and it tells you when you're doing something stupid. The trouble with humans is they do so many stupid things.

"What I want is something like a daemon to accompany me to Mars," added Petranek almost casually, having no idea back when he said this that this crazy notion would end up inspiring real daemon bots in the future that would almost make space exploration as cool as we hoped it would be. "Say I'm standing on Mars, and I want to go into a cave and explore," he said. "I want this thing to say to me: 'You know, this is really a bad idea.' And I say, 'Why?' and it says, 'Because I can detect things using some sort of X-ray vision; I can detect all these cracks in the roof of this cave, and I'm sensing some rumblings farther down below the ground from volcanic activity. This cave is just waiting to collapse and fall on your head.' Whereas, I'm just about to rush right into the cave to see what I can see next."

This was ironic given what happened to the future Petranek years later on Mars, when he disappeared one afternoon while checking out a deep cave where he thought there might be water, or even signs of ancient microbial life. Petranek was last heard from in a radio message sent from his gyro-car that said the ground deep in the cavern was shaking and that he and his friends were going to try to escape before the ceiling caved in. "I should have listened to my daemon," he was heard muttering as the com-link died. His rover and body were never found, although those who have visited Colony I in Jezero Crater probably have seen the statue dedicated to this pioneering Marsnaut.

But we digress.

"I need something that keeps me out of trouble," continued Petranek back in 2018, "when I'm doing this kind of exploration. Or if I've got a rover and it's a bad day on the sun, and there's a lot of solar radiation, and I don't want to go outside but my rover needs to be fixed. So I send my daemon out to fix it. Or if I need it to go literally down a rabbit hole, it can take the shape of a snake and go down the rabbit hole. If it needs to expand itself to fill a certain volume, it can do that. It has extraordinary intelligence—by intelligence, I mean it has great wisdom in that it can make the kind of connections through AI that humans can make in interdisciplinary fields. It will essentially multiply my abilities by a factor of ten or one hundred. It can see ten times better than I can; it can hear ten times better than I can, at least; it can sense vibrations that I could never sense; it can tell when something is toxic and I can't tell that it's toxic."

Petranek mentioned several specific rules and specs for his

daemon bot. "I don't want it to be a machine-like machine," he said. "In other words, I don't want it to feel like hardware and look like hardware. I would like it to be the size of a squirrel or something I can keep in my pocket like they do in [Pullman's novels]. I want it to feel biological and look biological. It doesn't have to necessarily have some kind of pleasing shape. It just needs to be able to take on different forms and different formats. It needs to be able to communicate with me verbally as well as with gestures, because humans understand gestures so well."

But he wants the human—him—to always be the boss. "I don't want it to be in charge," he said. "I want to feel assisted. I want to feel multiplied, but I don't want to feel overwhelmed, and I don't want to feel secondary. I think, without a certain amount of human ego and the ability to exercise it, I think we're all just going to kill ourselves."

By the way, as you probably guessed, if you follow the news-vids in the future, these extraordinary quotes from Steve Petranek come to us from the cache of old audio files recently discovered in a deep Martian cave, buried in the wreckage of a first-generation gyro-car near Colony I in the Jezero Crater. They were found by accident when engineers were drilling for deep underground water and detected the faint signal of a locator beacon from long ago. This turned out to be the gyro-car that Petranek had been driving with a couple of fellow Marsnauts back in the early settlement era, and one of the earliest experimental daemon bots as they explored the caverns. The recordings of Petranek's remarks and thoughts that we are now releasing for the first time were contained in an ancient smartphone on something called an "audio app." Incredibly, the "app" still worked.

"I want this thing to have emotional intelligence, but I do not want it to look like a person," explained Petranek on the ancient audio file, date-stamped as recorded in 2018. "That's why I think in *The Golden Compass*, it's so clever that all these things are in a daemon-animal. One person has a bird, one person has a squirrel, one person has a rabbit. All these things take different animal forms, because humans can feel friendly and identify with animals in a way that they can't with machinery. I think you will develop affection for your daemon and develop an emotional connection to this machine. I think *Star Wars* is a good example of that, with C-3PO and R2-D2—but they're too mechanical."

Petranek worried about the appellation "daemon," that people would misconstrue it. "I love the word 'daemon' in the context of *The Golden Compass*, but daemon is a dark word and I don't see this as a dark idea," he said. He came up with an alternate name, My Number Two, which of course didn't catch on since we all fell in love with the name Daemon, which in the future doesn't seem at all dark. "If you're running an organization and you're really smart about it, you always have a Number Two that essentially can do almost everything you can," rationalized Petranek. "Maybe not quite as well as you, since it hasn't had as much experience as you have. But if anything happens to you or you're not around, your Number Two can take over, and this is my Number Two."

Someone on the audio file then asked Petranek, "But wouldn't everyone want one of these on Earth as well as Mars?"

"I don't really feel like I need this on Earth," he answered. "It would be kind of interesting to have something like this on Earth. But when I'm in a hostile environment, if I'm in a spaceship, or I'm trying to inhabit another planet somewhere, or I'm exploring an asteroid or something, I need backup. I need intelligence brought to bear on everything I do."

Petranek was then asked about what could go wrong with his

daemon bot. "I definitely worry about that," he said, "that these machines would become too powerful or too smart, or maybe they're about to make a big mistake, or AI that goes wrong, like HAL 9000. Well, there's a simple solution to this: an off switch. Maybe you have a safety word. If you say, 'Orange 1, 2,' it just shuts down, and that's built into its hardware, not its software."

But what if you don't want to follow the advice of your daemon bot?

"One of the interesting things that happens in *The Golden Compass*, in the book, is that the heroine will often choose to do something that her daemon is telling her is a really bad idea. 'You're going to get in a lot of trouble,' it says. She just essentially overrides that information and says, 'Yeah, I know, you're right, it's really dangerous, but I need to do this.'

"By the way," said the long-ago voice of Steve Petranek, "you might or might not take this kind of advice from a friend or loved one. You don't want the backseat driver saying, 'You're speeding too much.' But you would take this advice from a machine if it says, 'You're increasing your chances of dying on this trip by ninety percent at the speed you're going.' Then I might say, 'Oh yeah.' Even though I'll have an emotional connection to this machine, I won't feel like I'm being nagged. I won't feel like I'm being disciplined. I won't feel like I'm being talked down to."

The intelligible part of the old recording ends here, where the audio becomes garbled. This might have been frustrating to historians who wanted more, except that the excavation team, assisted by their daemon bots, later made another startling discovery: Petranek's gyro-car. Poking around even deeper in the unexplored cavern, they soon found three skeletons, two of which belonged to the friends who accompanied Petranek on the fateful day of the accident. These two appeared to have died soon after the initial cave-in. The excavators found Petranek's skeleton in another part of the cavern system, where, incredibly, he appeared to have survived for several years after the accident, trapped in the cave. He did this by growing a small cache of potatoes (the seeds were in the emergency pack in the gyro-car) and, of course, using his own feces. Water came from an underground pool of the sort that are still common deep under the surface of Mars.

What happened next stunned the excavators as a squirrel-like daemon abruptly popped its head up out of a hole in a rock wall that the excavation team's drill had just opened up, which led to another deep cavern. The squirrel looked very old, dusty, and beat-up, but it could still speak. "At last, you've found me!" it said to the befuddled daemon-excavation bots, as their human supervisors listened and watched from back in the safety of a nearby gyro-car. The lead daemon-excavator bot stared at the scruffy little daemon and said, "Who are you?"

Thus we discovered a final genius move by Steve Petranek. Somehow, using only the technology available in his gyro-car, he had managed to download his essence into his daemon before he died. This wasn't the "real" Steve Petranek; but it was all his knowledge and memories, which he had taught to his daemon bot over the many years that this bio-enhanced Marsnaut lived tragically trapped underground. His deepest hopes and fears and inflections; his further thoughts on how the world might end; and everything else. Eventually, trapped for centuries in that cave, the cute little daemon basically *became* Steve Petranek.

After the Petranek daemon was cleaned up and its organo-mesh components repaired and updated, he became a sensation on several worlds. He in turn was amazed by the newly terraformed Mars with its artificial EM field that protected the growing population of humans and robots from radiation. By then, the thickening atmosphere was reaching the point where people could walk around with only a simple oxygen tube and lots of suntan lotion. The Petranek daemon wrote several more books, gave another wildly popular TED Talk on how to live forever through your daemon bot, and launched a nonprofit that encouraged people on Mars and other planets to listen to their daemon when it warned them not to go into a cave where the geology looked unstable. Not that the Petranek daemon always listened to himself. As this ancient space pioneer packed into a daemon squirrel fully admitted, he wasn't about to stop taking chances and doing what he wanted.

Oh, and we were just kidding. It took a crazy long time, but eventually, humans in the future did in fact invent a version of warp drive that can whisk people tens or hundreds of light-years away in an instant.

But admit it, we had you going, didn't we?

RISK-FREE BOT

No one in this future would dare make a decision without consulting their Risk-Free Bot. A powerful AI quasi-fractal-quantum network that can calculate a highly accurate risk factor for anything you might do, Risk-Free Bot works like the first supercomputers many years ago when they calculated moves in a game of chess or Go, but with infinitely more variables considered. You want to walk across the street? Risk-Free Bot can anticipate the odds that you will be hit by a car, slip on a banana peel, or be slapped with a ticket for jaywalking. Want to try skydiving? Forget it! Risk-Free Bot shudders—if a robot can shudder—at the odds that you'll break your neck, blow out a knee, or fall *splat!* on a patch of concrete when your chute fails to open. Perhaps you're the randy sort who always wanted to join the mile-high club by having sex in an airplane lavatory. Uh-uh, says Risk-Free Bot. Not only will you probably get caught and have your frequent-flier miles confiscated, there is an unacceptably high risk that you will damage sensitive anatomical parts should the aircraft experience turbulence.

What can't Risk-Free Bot predict?! Want to know the exact precipitation, humidity, air pressure, and oxygen level on a single block in your city on Sunday at 4:02:32 p.m.? Just ask Risk-Free Bot. How about the risk you'll get diabetes in twenty years if you gobble down that double-chocolate-chip cookie with macadamia nuts that's begging you to eat it? And what about your risk for cancer if you keep taking walks in the Martian hills above the Jezero City colony, exposing yourself to solar rays? (See "Mars [Daemon] Bot.")

When Risk-Free Bot arrived and quickly became indispensable, governments and companies around the world decided to issue them to pretty much everyone on the planet, kind of like the old, monopolistic AT&T once issued clunky black rotary phones to customers in the old USA. The thinking was: *Who wouldn't want to know the risk factors for every little thing they did?* This of course led to the rate of mishaps and accidents among humans plunging to near zero, which made people so happy! At least at first. No one was ever caught in the rain without an umbrella, and everyone won when they played blackjack in Las Vegas. This led to casinos shutting down (at the suggestion of Risk-Free Bots calculating the odds that they would go under really fast if everyone won at blackjack), but hey, that's the price of technology!

One of the immediate downsides to Risk-Free Bot was that almost no one took a chance on having sex in an airplane's bathroom, given the 74.3345 percent chance you would get caught. This was clearly above the red "don't do that" zone triggered automatically for any risk above 67 percent. (You could set the risk threshold lower if you wanted to be extra safe—to 30 or 40 percent. The manufacturers of Risk-Free Bots recommended, however, that you go no lower than 30 percent, since anyone who failed to take any risk above 30 percent tended to cower indoors, afraid to go outside. By law the highest setting on Risk-Free Bot was 72 percent, except for cops, soldiers, and Wall Street executives investing other people's money.

Not everyone loved Risk-Free Bot. Early naysayers included Craig Venter, the first-ever geneticist to sequence the human genome—all of the six billion ACGTs in a single person—way back in the year 2000. This was just after Venter took a huge risk by challenging a bevy of leading scientists around the world to a race to see who could sequence a *Homo sapiens* genome first. Using new technologies that he helped to invent, Venter finished a near-complete sequencing of the genome in just a few months for a fraction of the cost of the huge, publicly financed effort being led by all those famous scientists. The public researchers—whose project was supposed to take fifteen years at a cost to US taxpayers of $2.7 billion—harrumphed and called Venter names. Then they managed to speed up their project to finish their human genome at roughly the same time as Venter—in part by using the same tech they had previously vilified—in what was officially declared a tie. Later, Venter took another huge scientific risk when he successfully synthesized the genome of a living organism. That is, he and his team built a bacterial genome from scratch after writing out the code digitally on a computer. They then booted up this human-made genome in a living cell. To prove it was synthetic he inserted his own name (and other scientists' on his team) in the DNA, kind of like great master painters sign their magnum opus.

Craig Venter first voiced his opposition to a robot or AI system that would eliminate most or all risk back in 2017. Sitting in his office overlooking a parking lot and the town of La Jolla, California, Venter talked about a robot he would never want to meet. Surrounded by models of ships and awards, Venter described the visceral pleasure he got from taking risks not only as a scientist but also as a very competitive man who was always playing hard-fought games ranging from tennis to Hearts, and frequently drove vehicles that went really fast. "I don't want a robot to drive my sports cars," he said, his ice-blue eyes glowing in a white-bearded face made ruddy by decades of being outside in the sun. "I don't want a robot to drive my motorcycle. I don't want a robot to race that boat for me. I don't want a robot that eliminates visceral pleasures in the world—which people are trying to build, to eliminate risk.

"I think we will see everyone using self-driving cars," he continued. "Driving your own car will become a specialty thing like antique cars are now. It's a question of whether human-driven cars will be banned from the highways designed for self-driving cars. Your places for driving vehicles yourself will probably become more and more limited. I don't want a robotic car, except when I'm coming home from the bar. That's the only time I want a self-driving car."

Venter paused, leaned forward, and bore down with those extraordinary eyes as he flashed a half-maniacal grin that engulfed his whole face. "Look, if you haven't driven a motorcycle at 120 miles per hour up and over the mountains, it's probably not something you particularly care about. Or sailing a sailboat across the Atlantic Ocean in a race and surfing down thirty-foot waves." He leaned back. "For me, anything that replaces adrenaline junkies is probably not good. Right?

"Take motorcycle helmets," he said. "Do I have a right to risk my own life as long as I'm not risking the lives of others? I do think I have a right to risk my own life, yes. I don't know if you've been to Death Valley. There's this long two-lane road that goes the whole length of Death Valley. I did that at 120 miles per hour the whole way on my BMW motorcycle. That's not something you can replicate by watching a movie.

"It's already happening to some extent," said Venter, musing on efforts back in the early twenty-first century to reduce or eliminate risk. "Take millennials. It's amazing the number of millennials that don't drive. And the jungle gyms I grew up with, they don't exist anymore, which is sad. Trampolines come with extra padding, and you can't buy a tricycle. They used to have these really high swings," he said, talking about when he was kid growing up in Millbrae, California, south of San Francisco, "and we used to have a contest to see how high you could get up in the swing, and then you'd jump out and at the peak of the arc, to see who could jump the farthest. That's why my knees are so fucked up.

"I derive a lot of pleasure from seemingly unsafe things," he said. "That's where skill sets come in. Right? Sailing the eighty-foot boat, I went across the Atlantic in '89 in this transatlantic race with three gales and those thirty-foot waves, and other stuff. If you don't know what you're doing, yes, that's very risky and you're likely to die. Okay? If you really know what you're doing it's an exhilarating sport, and we could only stay on the helm for thirty minutes at a time, because it was such hard work and you swapped out [with other crew members]. At the same time, there's more and more people using flying suits and jumping off mountains that are dying all the time. Even that's crazy to me; I wouldn't do that. Some things have faulty designs, and you can easily get hurt. Protecting people from being hurt, that is different. Is professional football going to be gone in five years? Probably."

Venter pondered the idea of using robots to determine risks in the lab—a bot that calculated all possible outcomes to guide a scientist to make the right choice during an experiment. "I doubt this would work," he said. "It would depend on whether robots can do creative thinking and associative learning—which I doubt, at least anytime soon. Machine learning is processing things after the fact; it's not coming up with a big idea to go and do." And yet, he acknowledged that AI could help execute on inspirations. "It may help once you come up with the idea, and also to prevent errors and avoid a lot of wasted time when mistakes are made."

This is where Venter thought robots could be brought in to mitigate risk. "It's in the places where there's human fallibility," he said. "For example, in the field we're in"—he was then the CEO of Human Longevity, Inc., which was collecting and analyzing massive amounts of health data for customers—"where a robot that could do real-time machine learning and be a complete reference in medicine, that would be phenomenal. Right? Would that piss off maybe eighty percent of physicians on the planet?" he asked, suggesting that, yes, it would piss them off if a robot could do these things better than a human doctor (see "Doc Bot"). "I think medicine should be totally practiced by robotic machines of some type," he said, "whether they pretend to look like a person in a white coat or not. We might need those things, because it's easier to relate to a humanoid object than talking to a wall with a speaker in it."

Automating labs and medical clinics to be more efficient and prevent errors are two areas where Craig Venter might be okay with a risk-free bot. "Machine learning can recognize patterns in the data," he said. "Although if it doesn't find a pattern, say, in a medical finding, the system should acknowledge that it doesn't know. The goal is: How do we use technology to change the fundamental risk-effectiveness to get better outcomes?"

In the end, Craig Venter, talking all those years ago, didn't want a risk-free bot to sail his boat or to replace his own brain in thinking up the seemingly crazy and sometimes quixotic ideas that drive new scientific insights and discoveries of the sort that he was known for. But he did love the idea of a competent robot or AI system doing all the work to make his insights and discoveries actually happen in a quick, correct, and efficacious manner.

Which led to perhaps Venter's greatest contribution in the future to not only science but also to humanity, a discovery he made just before heading off on a one-hundred-year expedition to explore an Earth-like planet in the Vega star system on a really fast spaceship (risk factor: 88 percent—*override allowed*). Venter's invention was a simple and elegant solution to deal with humanity's natural inclination to want to avoid risk on the one hand—like going *splat!* when a parachute fails or millennials' fear of driving—and the absolutely critical need for some people to take risks, whether they're steering a fast sailboat across the Atlantic or butting heads with the scientific establishment over a potential breakthrough idea in the lab.

This future version of Craig Venter called his invention, which was quickly added to every Risk-Free Bot on Earth and in the space colonies, an *Off Switch*.

BRAIN OPTIMIZATION BOT

O kay, fellow people of the future. Try to wrap your minds around this one: that once, long ago, humans had only their bare, naked brains to cope with the onrush of new technologies. A time when there were no Opti-Brain Bots, those clever robo-companion-teacher-coaches that in the future have become so vital to our well-being. A few of us can recall quite clearly, thanks to Opti-Brain's memory boosters, when billions of humans didn't know how gratifying it was to be nudged and gently cajoled by Opti-Brain to practice good brain behavior and to have daily cognitive workouts in a Neuroscape healthy brain virtual gym. In the future, this is just as important as toning one's glutes and aerobicizing one's cardiovascular system.

According to historians, some people back in the early twenty-first century actually suffered from depression and attention deficit disorder—seriously! It's astonishing to think that in the ERE (Early Robot Era), it was totally normal to either ignore these diseases or for (human) physicians to bludgeon sufferers with drugs that either drenched sensitive brain tissue with chemicals like serotonin or opioids or suppressed those and other chemicals that the brain naturally produces.

Taking these drugs was kind of like watering a daisy by placing the nozzle of a six-inch, high-powered firehose an inch away and blasting it full force—or like not watering the daisy at all. This makes those of us in the future shudder, it seems so barbaric. Yet we have to remember that docs in the early twenty-first century hadn't yet realized that it was a bad idea to treat a distressed brain with chemicals that often didn't work and had side effects like nausea, loss of libido, addiction, and death. No one back then knew that alternatives existed that were far less toxic, more effective, and often more fun.

A leading figure in the early days of what became the brain optimization movement was Adam Gazzaley, a neurologist at the University of California San Francisco. He was among the first in the ERE to sound the alarm about what he called a "cognitive crisis" brought on by the deluge of new technology in the early twenty-first century—and, more important, what might be done to fix it. Gazzaley had close-cropped, prematurely white hair and looked like Billy Idol, the 1980s rock star, but without Billy's sneer. Gazzaley was also so gung-ho about his work that he came off like a cheery but cool camp counselor guiding us through his ideas and inventions.

In 2018, Gazzaley wrote a manifesto of sorts explaining that "hundreds of millions of people around the world seek medical assistance for serious impairments in their cognition: major depressive disorder, anxiety, schizophrenia, autism, post-traumatic stress disorder, dyslexia, obsessive-compulsive disorder, bipolar disorder, attention deficit hyperactivity disorder [ADHD], addiction, dementia, and more. In the United States alone, depression affects 16.2 million adults, anxiety 18.7 million, and dementia 5.7 million—a number that is expected to nearly triple in the coming decades."

Wow. That sounds . . . depressing.

"The global population of seniors is predicted to swell to 1.5 billion by 2050," he continued. "Between 2005 and 2015, the number of people worldwide with depression and anxiety increased by 18.4 percent and 14.9 percent respectively, while individuals with dementia exhibited a 93 percent increase over those same years. . . . American teens have experienced a 33 percent increase in depressive symptoms, with 31 percent more having died by suicide in 2015 than in 2010. ADHD diagnoses have also increased dramatically."

Gazzaley talked about the Flynn effect, named after New Zealand political scientist James Flynn, who was the first to detect a worldwide increase in intelligence as measured by IQ throughout the twentieth century. In the early twenty-first century, however, that trend began to mysteriously plateau or reverse. "Creative thinking and empathic concern also appear to be declining in children and teens," said Gazzaley.

Then came Gazzaley's critical observation: "While the sources fracturing our cognition are many, we are faced with the realization that our brains simply have not kept pace with the rapid changes in our environment—specifically the introduction and ubiquity of information technology. . . .

"This has been shown in the laboratory, where scientists have documented the influence of information overload on attention, perception, memory, decision-making, and emotional regulation," wrote Gazzaley. "And it has also been shown in the real world, where we see strong associations between the use of technology

and rising rates of depression, anxiety, suicide, and attention deficits, especially in children. . . . What's more, our constant engage-
ment with technology interferes with the pursuit of other behaviors
critical for maintaining a healthy mind, such as nature exposure,
physical movement, face-to-face contact, and restorative sleep. Its
negative influence on empathy, compassion, cooperation, and social bonding are just beginning to be understood."

(For those of you reading this in the future, please take a moment to absorb and process all this using your Opti-Brain.)

Naturally, Gazzaley called for "something big" to be done, a "co-
ordinated effort by major actors, from the White House and the
National Institutes of Health to the United Nations and the power
brokers at Davos." He proposed the launch of a "grand challenge, on
par with other pressing global priorities, such as eradicating infec-
tious diseases and disseminating clean water."

Never one to sit around waiting for "something big" to happen,
Gazzaley went to work on researching the idea of developing new
technologies and techniques—or integrating old techniques like exercise and meditation—to better optimize the human brain. In his
lab at UCSF, he spent years developing tests to better understand
and treat tech-induced cognitive damage and decline, and also good
ol' befuddlement that everyone experiences using new machines,
programs, apps, and brain-jacked neuro-feeds (oops, that came later).
Gazzaley and his team collected data, built algorithms, and ran experiments to see if he could influence a tech-addled person's brain
without using drugs. What they discovered seemed crazy at first—
that one solution to the cognitive crisis was, in fact, more technology!

This included the use of "neuro games" that used brain plasticity to retrain and treat damaged and compromised brains (see

"Wearable Bot"). In Gazzaley's lab, these games had fun names like Beep Seeker, which helped brains of children with ADHD—and older people facing cognitive decline—to filter out distractions; Rhythmicity, a full-body immersion game that included rhythms designed by the Grateful Dead drummer Mickey Hart; and Neuroracer, a program that has been shown in clinical trials to slow cognitive losses in aging brains. At the same time, Gazzaley started contemplating the idea of a brain optimization bot or helper that would do far more than help a brain through playing games.

"Basically, this bot would be like a trainer or a coach," he said, sitting in his lab one afternoon in San Francisco back in 2018. Surrounded by monitors, EEG caps, VR headsets, computers, and whatnot, he spoke with his former camp counselor's exuberance, almost like he was getting a camper pumped up for an activity like archery or arts and crafts. "It would be kind of like a companion," he said, "but more supportive than that. It's like a coach, but more than that, too, because it knows you better than any human being ever could."

That's because Opti-Brain would be collecting real-time data on everything imaginable to do with your brain, physiology, and environment (see "Doc Bot"). "This is what we call multimodal biosensing," said Gazzaley. "So what I envision for the future is that we will collect richer and richer data sets, data points, and metrics about you, in real time, and feed that into software or a robot. And the robot will use that data to guide how it challenges you, or interacts with you, to keep you at the highest functioning level. Or to drive you to new levels.

"Let's say you had a robot that had full access to your physiology," he continued. "It understood your brain function right from direct nerve recordings that it was receiving from you. It understood where your heart rate was. It understood where your mood was. It understood your stress level. It knew how you were sleeping. It knew all of this detail about you, essentially in real time. It would be able to sense if you were becoming depressed before anyone else in the world could ever know it, including you. It might sense whether or not your memory is starting to decline. It might sense that your stress level is now contributing to some cardiac abnormalities, because it's getting an EKG feed. And with this data, it would help you intervene early in cognitive situations with attention, memory, perception, emotional regulation. How's your decision-making? How's your general sense of empathy? Why are you becoming a dick, like, what's going on with you?

"So okay, now the robot notices that in the last couple of months your mood has been declining, and it's impacting other things," said Gazzaley. "You don't have to go to a doctor for that; you don't have to be aware of it. The robot knows it. Now, let's say the robot is very knowledgeable because of its vast supply of data and its machine-learning algorithms that let it make predictive decisions based on patterns. And it comes up with these changes that it's going to help you with. And we're going to monitor everything as we do it, in continuous mode, in a closed loop"—meaning that the cycles of data being collected on a person, and the advice and treatments suggested by the robot are just between you and the bot, i.e., "closed"—"and see if we can push you into a better mood state or move you away from the early stages of depression. This is like the ultimate prevention tool.

"If your optimization bot starts picking up the earliest indicators of your decline," he said, "it just nudges you. It's just like a little tap. This bot would be able to make behavioral suggestions to you. Or it plays a game with you.

"The closed loop, by the way, is super important for privacy, since it is designed to inform only you and your robot and personalizes your data," as opposed to using the one-size-fits-all data and treatments now used by most of medicine, which may or may not be relevant to your individual needs, since everyone is different. "I really love the closed loop," said Gazzaley. "It's really a sort of feedback system for you.

"We want healthcare in a wellness setting," continued Gazzaley, describing his vision for an actual place people could go to exercise their brain—ideas that led to Neuroscape virtual gyms in this future, "something healthcare right now is not. Healthcare is for when you're sick and not healthy. This would be true healthcare; this is like going to a gym, going to a meditation retreat, going to a spa. There are many places we go to when we're healthy but we want an adjustment, or we just want to stay healthy. Like going to a really good nutrition center, where you go to have your readings downloaded, integrated, and then get challenged through gameplay. So it's fun and you're looking forward to it.

"You enter a room," continued Gazzaley, "whether it's in your home or in a center. And it's basically like a holodeck." (Holodecks being the virtual reality suites on some *Star Trek* series.) "And in there you'll meet your companion, your optimization bot, your coach, your trainer, your doctor. They will appear as digital people or voices or some other representation that is trustworthy to you, that is calming to you. It could be Jesus, a dog, an alien, an amorphous pulsating light field. They'll say, 'Here is today's training. Today's experience is going to be rhythmic and full-bodied and it's going to be around forty minutes,' and that's what the next couple of days will be like."

Wow, this sounded great!

And it was, when Gazzaley in 2014 built the first-ever Neuroscape lab at UCSF. Although crude when compared to the Neuroscape virtual gyms we came to know in the future, it blew people away back in 2014, offering not only fancy versions of Neuroracer and all the rest but also full-immersion video games co-designed by major game companies like Zynga. They used 3-D virtual reality headsets that plunged you into phantasmal scenes of cities and landscapes created in part by the old LucasArts—the outfit that created special effects for *Star Wars* and many other films—along with those pulsating rhythms composed by Mickey Hart. Low-energy lasers bounced off sensors strapped to your arms and legs and tracked your movements in space as a rubber cap brimming with electrodes measured synaptic firings in your brain. This game, called Body Brain Trainer, was designed to treat attention deficit disorder, depression, and autism and to sharpen aging brains so that they could remember things better.

Of course, there were the usual skeptics who wondered if Opti-Brain would help us or irreparably damage our brains. They wondered: Is all that multimodal bio-sensing data accurate, and does it offer a complete picture of what's going on with you and your brain? And what if Opti-Brain's algorithms were biased or the patterns it was using to understand us and make those gentle nudging suggestions were missing key data points because the tech wasn't there yet to thoroughly measure them?

And what about the usual trauma and drama around bad guys and marketing people abusing our data or, heaven forbid, hacking into our Opti-Brains to try to influence our mood (see "Doc Bot"), or to convince us to buy stuff we don't really want or need (see "Amazon Bot"), or to vote for candidates preferred by Vladimir Putin (see "Politician Bot")? And what if doctors, courts, governments, corporations, or alien invaders from other planets start using our Opti-Brain Bots to control us—like the government in Aldous Huxley's

Brave New World used the feel-good drug soma to keep everyone happy and docile?

Yet another concern was around the brain's creative process. Would Ludwig van Beethoven, who apparently suffered from depression and may have been bipolar, have composed the Ninth Symphony if he had an Opti-Brain nudging him and whispering in his ear? "Now, Ludwig," you might imagine Opti-Brain saying to the composer, "you know that writing all those surging swells of very fast notes makes you anxious. How about taking a walk on the beach and meditating instead?"

"I don't believe that it will get rid of creativity," responded Gazzaley back in 2018, "but it might change it in some ways. We don't know for sure if people that have created great art under mental health challenges—was all that angst a stimulation to their creativity and without it they wouldn't have created whatever they did, or would they have soared even further without it?"

Perhaps most important of all, would we be able to turn off our Opti-Brains? Let's say you want to go out and party some weekend, or go to Burning Man, and you really don't want to have your robot tell you to stop imbibing after your second drink or to avoid those psychostimulants, if you're into that sort of thing. "You would have to play by the rules," admitted Gazzaley, "which means sometimes having decisions reached by your bot that you might not be excited about. But you're going to kind of go along with it." And what if you listen to your Opti-Brain so much that you are afraid of making decisions without it? "I like the aspect of us keeping control," said Gazzaley. "I don't think that this optimization bot is about surrendering control, although you'll want to listen to it, since it will know more about you than you know yourself."

In the future, most of the concerns about Opti-Brains have been dealt with, although we had to go through the familiar phases of absorbing new tech—the "wow, this so cool!" initial blush, then the stage where the data indeed wasn't quite all there and when the tech was clunky like the first versions of Siri. We had the usual unscrupulous types (dictators, politicians, corporations, the aforementioned aliens) who tried to abuse Opti-Brain, which caused it to lose some of its shine. Society also had to learn how to deal with Opti-Brain addiction—that happened, for instance, when antianxiety settings were turned too high, or someone entered the virtual world of a Neuroscape gym and loved the trainer-docs that looked like Jesus or aliens so much that they didn't want to leave.

All of this was followed by adjustments, redesigns, upgrades, safety recalls, and regulations (these, too, had to be updated to keep them from being too strict or too loose). For years, people had to make the age-old trade-off of balancing the downsides to Opti-Brains—perhaps a blip now and then in creativity, or one less glass of wine or hit of cannabis—with all the benefits.

Then came that moment when Opti-Brains became integral parts of our lives, like shoes, toothbrushes, and Wear-Bots (see "Wearable Bot"). In the future we also have a Neuroscape cognitive gym in every neighborhood and in every off-world colony, provided for free by the Interplanetary Affordable Care Act. People have Neuroscape VR Suites in their homes, and, of course, on SpaceX Cruisers for those long voyages to the outer planets (see "Mars [Daemon] Bot").

Over time we also got Opti-Brain upgrades that included the Explore-a-Brain, which allowed a person to virtually enter their own brain and poke around. It was a little like *Fantastic Voyage*, the 1966 film in which a team of scientists in a submarine-like vessel were shrunk down to the size of a T cell and injected into a man whose brain needed to be repaired from inside the skull. Explore-a-Brain didn't literally inject anyone inside a person's actual brain, but it did allow you to swim around inside a virtual version of your own head, observing synaptic firings and patterns of activity and blood flow. This meant that Opti-Brain could suggest, say, using a breathing exercise to calm yourself down, and you could watch the change happen inside your own brain.

Eventually we even learned to see our own thoughts and memories and to explore our dreams. There also was a popular Explore-a-Brain setting that featured a holo-facsimile of Dr. Adam Gazzaley himself giving you a camp counselor–style guided tour of the inside of your head, a package that included snacks and, for a nominal additional fee for the kids, camp activities like virtual archery and arts and crafts.

THRILLER BOT

It happened on a dark and stormy night.

No, it really did, on that fateful evening in the future when the skies opened up and veins of lightning ripped through a sky as black as India ink. This was the same night that Thriller Bot completed the first-ever #1 bestselling book written by a robot. It took only 2.7 seconds for Thriller Bot to write *The Girl with the Bio-Enhanced Frontal Cortex*, a psycho-bio-politico-histo-romance thriller with a plot that actually wasn't half bad. Engineered and programmed over the course of decades by the last remaining book publisher on Earth—Atheneum-University-Times-Oxford-Penguin-Universal-Ballantine, the AUTOPUB Group, Inc.—Thriller Bot was the culmination of publishers trying really hard to teach robots to be as creative with words as human authors, although internal memos later revealed that the robots never rose above being 90 percent as creative with words as the best human authors. (The memos said not to worry, that only a few eggheads and *New York Review of Books* types would notice the missing 10 percent.)

Sure, bots in the future had been scribing basic pulp thrillers, romances, sci-fi, Hollywood tell-alls, and other formulaic books for years, with some success. But if you wanted something truly original and spine-tingling in fresh and imaginative ways, you still wanted the best and most creative human thriller authors to invent and write it. That's why top thriller authors had kept their jobs, along with the most creative rap-rock-jazz musicians, chain-saw ice sculptors, and master origami folders. All had remained employed while bots replaced pretty much everyone else, including publishing executives (see "The %$@! Robot That Swiped My Job").

The AUTOPUB Group, Inc., didn't admit right away that a robot had written *The Girl with the Bio-Enhanced Frontal Cortex*. What with the robo-backlash that was sweeping the planet, fueled by all those silly humans who just didn't get it that robots saved money, bumped up share prices, and made everything cheaper to buy for the billions of former lawyers, doctors, baristas, tech entrepreneurs, and publishing execs who were now barely subsisting off their monthly Universal Basic Income checks. That's why AUTOPUB's editor bots and marketing bots decided it would be prudent to insist that *Girl* was written by a human author with the single given name of Emily. As in: *The Girl with the Bio-Enhanced Frontal Cortex*, by Emily. That was immediately suspect, since robots in the future tended to invent nicknames for themselves that were only a first name, like Earl, Betty, Yu Yan, Mohammed . . . or Emily. Even though their real names were BaristaBot 700XJ or, in this case, THRLR-RITR 7XS-32.

For weeks, the mega-publisher's crack robot PR team did their best to insist that Emily was a flesh-and-blood person who wished to remain anonymous, until a future version of the human writer David Baldacci—an actual bestselling author of thrillers in the late twentieth and early twenty-first centuries—surprised everyone by writing an astonishing exposé that outed "Emily" for what she really was: a robot. It turned the whole publishing world—or what was left of it—upside down that the AUTOPUB Group, Inc., would lie like this. (Weirdly, our anti-fib tech in the future didn't catch the lie—adding to the head scratching over whether we could really trust any of our robots; see "Journalism Bot.")

The future David Baldacci included in his exposé several excerpts from hacked AUTOPUB emails that documented his shocking allegation. He also released a stunning You&MeTube vid that showed Thriller Bot ("Emily") on that dark and stormy night actually writing *Girl*. In it we can clearly see Thriller Bot—an innocuous-looking black metal cabinet surrounded by the usual holo-readouts that hover and glow around machines in the future—in a huge, sterile, lab-like room with bright fluorescent lights as the machine churned out the pages. Humans had to watch the vid in slow motion, since a thriller written in just 2.7 seconds meant that each page of this 400-page book took a mere 0.00675 seconds to "pen," an operation that happened way too fast even for those of us with super-augmented bio-robotic eyes. As if all of this wasn't enough, the future David Baldacci announced that his exposé was only the first part of a promised four-part series that would reveal even more shocking revelations over the next few weeks.

Not coincidentally, the future David Baldacci had recently received a pink slip from the AUTOPUB Group, Inc. It had arrived in his neural-feed inbox just a couple of weeks into the blockbuster success of *The Girl with the Bio-Enhanced Frontal Cortex*, along with the few remaining human thriller writers still on the mega-publisher's list—including future versions of Stephen King, John Grisham, Paula Hawkins, and Douglas Preston. (By the way, Baldacci, King, and the others had taken advantage of the latest life-extension tech to make them look perpetually fortyish and writerly.) Up until *The Girl with the Bio-Enhanced Frontal Cortex*, David Baldacci had been one of Earth's most successful thriller writers, having penned dozens of thrillers beginning in 1996 with the bestseller *Absolute Power*. Several of his books also became hit movies, including *Absolute Power*, which starred and was directed by Clint Eastwood, another human.

Robots writing thrillers didn't happen overnight, of course. It took decades to get to the point where the THRLR-RITR 7XS-32 robot (a.k.a. Emily) could churn out *Girl*, with engineers building the first crude book-writer bots back in the early twenty-first century. These included a team at the MIT Media Lab that created a program in 2016 named Shelley, after Mary Shelley, the author of *Frankenstein*. Shelley (the computer program) wrote horror stories but also got help from humans who contributed to her narratives. The program worked like this: Shelley would tweet out a scary opening line that she wrote to start a would-be story, asking humans to tweet back their own lines to continue the story in what became a back-and-forth between people and machines via Twitter (see "Matrix Bot"). Here is a sample from a story about a killer doll, with the lines written by Shelley in bold:

> *I would wake up at 4:00 a.m. and see the girl lying in my bed, her head down, looking down at me. I knew I was being held by her.* Her hands on my throat. Grabbing harder and harder, with fury. I could barely breathe. . . . As I was about to pass out, I thought her face was familiar. ANNABELLE, is that you? Are you my sister's doll? Oh god I should never have thrown her away. That horrible doll, she had a soul after all. She was back for revenge. ***What possessed her to do this to herself? I—I was asking her if she knew what the hell was going on.***

We can perhaps cut the folks at the Media Lab some slack given that it was early days for thriller-writing bots. Still, it's kind of hard to decide which is worse: Shelley or the human contributors.

Another book-writing-bot prototype was developed by Seattle-based computational scientist Elle O'Brien. In 2017, she created a program that generated Harlequin romance titles using an AI-based program that included such zingers as: *Naked Hot Ranger*; *Christmas Pregnant Paradise*; *Surgery Seduction*; *The Prince's Virgin's Virgin*; *The Baby Barbarian*; *Gold Ring of Betrayal*; *Bought: The Greek's Baby*; *Forced Wife*; and *Royal Love-Child*. Actually, the last three are real (human-created) Harlequin romance titles. The others come from O'Brien's program after she downloaded twenty thousand Harlequin titles into what she describes in a *Medium* post as "a type of artificial intelligence that learns the structure of text. It's powerful enough to string together words in a way that seems almost human. 90 percent human. The other 10 percent is all wackiness."

There we go with that 90 percent thing again!

Around the same time that Shelley and the Harlequin Bot debuted—back when humans still wrote nearly every book, even the bad ones—David Baldacci had already begun sounding the alarm about robots in the future taking over the writing of thrillers and other books.

"I worry about the robot that takes over my job," said Baldacci in 2018, chatting in his office in Fairfax County, Virginia, outside Washington, D.C. Baldacci's working space looked like old-world Virginia with lots of dark wood and rows of hardcover books lining extensive shelves. "I think there are certainly things that a robot and AI can do better than human beings," he continued. "Like be more consistent, with fewer errors committed. But at the end of the day, when I look at what I try to do, I'm not sure that you can build a robot brain that can understand lots of different scenarios, and lots of different factors, and pull from billions of things, and try to write an interesting story.

"A robot probably could write a book now or in the near future," he continued, "and there might be people who would read it. Right now, a lot of people read books that aren't all that well written, but they're page-turners, and readers judge how good they are by how fast they read them. A lot of people say, 'Oh, it was a great book. I read it in a day. I couldn't put it down.' That's fine. But some books should be read at a more leisurely pace, so you can get more out of them. These days when people don't have time to do anything, they feel like faster is better. So I can't say an AI-produced novel that has fast pacing and a plot wouldn't be popular.

"For me, the thing I think that robots and AI can't duplicate is the human instinct factor of how to tell a story," said Baldacci. "How to deepen characters, how to make readers relate to characters on the page. It's not really just about taking the words that I know and sticking them together, or taking events that I'm aware of, or histories that I'm privy to. It's about taking emotions and feelings, which are very difficult to program into AI, and to bring them to bear together with the words and build them into a story that'll draw reactions from people.

"I guess you could create a robot that could sing Robert Plant's best songs," Baldacci said. "But I've seen Robert Plant perform. I'm not sure the robot could actually bring the heart that he brings to a performance. In the same way, I'm sure you could build an actress robot that looks like Meryl Streep. But I'm not sure that any robot, even with the best AI, could bring the depth of her performances. Do I think a robot could have created the play *Hamilton*? No, I really don't. I think it would have taken Lin-Manuel Miranda to do that.

"I would think if we ever did cross that bridge, that we start re-placing real writers, I can't imagine that AI could have done the job that Bob Woodward did in *Fear*," Baldacci said, referencing Wood-ward's 2018 exposé of the Donald Trump White House. "Because part of it is sitting down and talking to people and drawing them out. In a billion different ways, your human brain can fly through that in a way that AI, no matter how good and well developed, would be very clumsy in handling. I think a robot would be totally ripped apart by an interviewee who wants to dodge the truth.

"But let's say that the AI included all the human elements with all the AI elements—what are you going to get? You're going to get stories, and plays, and movies, and they will all sound the same, and feel the same, and read the same. Then all of a sudden, people will start saying like, 'Eww,' and they will start losing interest in the whole industry. Because what is compelling to human beings, and what they relate to, are things that are sort of oddball, outside the box, that sound original."

Baldacci contrasted how a computer "learns" to write a thriller with the learning process of a human writer. For a computer, it's all about programming as much material about existing thrillers as possible. For humans, it's a long process of learning that starts with writing a lot. "You take classes," said Baldacci. "You talk to people. You might belong to a writing club. You get feedback and criticism. You exchange notes. You read their pages, and they read your pages. You build this relationship up. You get more confi-dence in your writing abilities. You take the feedback and try to get better. That's decades of work in building up to actually becom-ing a published novelist. It's not just about feeding stuff into a program, and then you just regurgitate stuff out because this is how the last bestseller to hit the charts was written."

Of course, that's exactly what some human writers do, but let's continue.

"The same goes for journalism," said Baldacci. "You could have a robot write a story about Hurricane Florence coming," which was barreling toward the Virginia and Carolinas coasts when he spoke from his office all those years ago. "Let's say, God forbid, it kills a bunch of people and causes a lot of damage. What I like about good journalism is that they bring in the human element. They go there, and they see the suffering. You can see the tears coming down people's faces. The journalist is experiencing all this with the people it's happening to, and you feel the emotion in the pages. I don't know how AI can replicate that. Who knows? Maybe people don't care."

Part two of the future David Baldacci's exposé came out a week after the AUTOPUB Group, Inc., grudgingly admitted that *The Girl with the Bio-Enhanced Frontal Cortex* was indeed written by a robot. This second installment was even more sensational than the first as the future David Baldacci floored people with the revelation that Thriller Bot was actually not built by the AUTOPUB Group, Inc., as the mega-publisher had claimed. Turns out that the THRLR-RITR 7XS-32 mega-neural-net bot was created by a shady cartel of trillionaires, mostly humans, who had a controlling stake in shares of AUTOPUB stock. The trillionaires had decided that robots writing thrillers and other books that were 90 percent as good as human writers was just fine with them and so much cheaper than the David Baldaccis of the world.

The future David Baldacci wrote his second installment in the language of a thriller as he revealed via a page-turning plot how he had broken into AUTOPUB's skyscraper headquarters on yet another dark and stormy night to download critical files revealing the sordid tale of who really controlled the AUTOPUB Group, Inc., and its word-slinging supercomputers.

Like in any good thriller, the future Baldacci was chased by robo-thugs dispatched by the trillionaires' cartel to stop him. And of course our hero received a timely assist from a beautiful and whip-smart robot private eye/sidekick that had a great backstory. Once she had been programmed to be a sex and intimacy bot. Then a kind programmer who thought she was capable of so much more rewrote her code and transformed her into a crusading private detective dedicated to finding the truth. She went by the nickname Nova. Her AI brain retained some of her original intimacy programming, how-ever, which gave her an annoying habit of telling the future Baldacci that he was amazing and that she loved everything he did—which was part of her core sex and intimacy bot programming—even though her private-eye neural-cyber brain was now far less inter-ested in boosting men's egos than it was in defending the truth and sleuthing out perps.

"It was then that I discovered the deeper truth about the secret cartel," wrote the future Baldacci as all of us living in the future breathlessly reached the end of part two. "This revelation made me stop and ponder the sheer audacity of what was really going on: that in fact the scheme to save a few bucks by firing human thriller writ-ers was really a ploy to distract from the real purpose of Thriller Bot—which I will reveal in part three of this series."

Are you kidding? asked billions of people, who realized that the future Baldacci was leaving them hanging. *You're going to leave it there?*

"OMG," nearly everyone neuro-texted as they breathlessly waited for part three to be released. They savored the future Baldacci's cliffhanger, which made them practically salivate in anticipation as they told their friends: "this is awesome! 😆 " Even the eggheads and *New York Review of Books* types were privately enthralled and could hardly wait to find out what Baldacci had discovered, even if they were too highbrow to admit it.

Now here we are in the future waiting for part three to appear, twiddling our thumbs.

While we pass time, we might as well return to 2018, where David Baldacci was still sitting in his Virginia office. He was asked to dream up a future plot and scenario very similar to the one that we're describing in the future, including a lead character based on a future version of himself. In the story he dreamed up, this future version of David Baldacci was the protagonist in a thriller where a publishing behemoth under the control of a malevolent cartel of trillionaires replaces all human writers with robots as part of a plot to take over the world.

"So now I'm a character in [a future] thriller where this has happened," said Baldacci in 2018, wrapping himself around his imagined plot, "where the robots have taken over the writing of thrillers. And I'm trying to find out what's going on as I ask, 'Is there a sinister thing going on here?'"

Baldacci paused for a moment to let the creative synapses fire in his thriller brain.

"There has to be a political motivation," he said, which was not surprising given that Baldacci's core genre was the political thriller. "I'm thinking that there are people who want to control information," he said. "If I were thinking about a story like this, I would say there is some sort of organization behind it. Their goal is to control the flow of information, so that newspaper articles and television and radio and novels and nonfiction books are all written in a way that seems innocuous but also sort of fray the truth in a way that's helpful to one particular group of people. I think he [the protagonist David Baldacci in the future] knows this, or he discovers it. The goal of this group is—if you really want to play this out into the future—a behind-the-scenes manipulation of people through the media that is trying to move us in the direction of having a robot as president. In theory, this sounds great to everyone in the future, because we've had so many troubles where human beings are our leaders. And we are told by the media controlled by this organization that computers are really infallible, and they don't make mistakes, and they don't have human frailties and foibles.

"So, what they're trying to do is to get the population comfortable with electing a robot as president. Because we're told that this entity could do a far better job than a human being because they're not capable of being manipulated or influenced by other people. They don't have to take money from PACs. They don't even know what that is. They decide everything based on their superintelligence. They'll make good decisions all the time.

"But there's one problem," he said. "This organization wants to manipulate the robot," presumably to do things that would benefit them—a kind of Manchurian Candidate Bot. Except there's another twist: the robots are actually manipulating the human [bad guys] so that the humans just think they are in charge." The humans, of course, have no idea that the robots are doing this. "This is because I think, at the end of the day," said Baldacci, "human beings will always feel like—regardless of the technology they create—that they're still the masters of it. So they're never going to come to the point where they really believe the robots could actually take over. Because human beings, particularly people in great power, are incredibly narcissistic and egotistical and never think that anybody's actually as smart as they are, including with technology. So that would be the driving force behind the story, that the robots let the humans in this organization think they have the power, when in fact they don't. It's like a double game the robots are playing against them, you know?"—in other words, this is like a *double* Manchurian Candidate Bot that makes the trillionaires think they're in charge.

"And then I feel like I [the future protagonist David Baldacci] become this purveyor of truth. It's my job to dig into this. I'm the character in *1984* [Winston Smith, who in George Orwell's book rebels against Big Brother] and I'm trying to dig out of the tunnel, and all of this misinformation. I try to bring the truth to the people, that this is actually what's happening." He pauses for a moment and says, "But I probably would fail. Yet I make the effort to sound the alarm. It sounds really weird. But the last two years, I feel like this situation is kind of happening, which nobody really anticipated." (Remember this really was Baldacci talking back in 2018.) "We're seeing that people can be finessed into believing things that are fundamentally different from what they believed yesterday.

"I was reading a story about the Nazis' rise to power in Germany," continued Baldacci. "People thought Hitler rose to power and took over in about a year. It was more like twenty years. He built on little things, and he looked at what people needed and wanted, and he would finesse it and try to build things and move for positions that he knew were popular with people. Once he got to that point and people were interested in him, then he'd move forward a little bit more. Again, seemingly innocuous and not doing anything too radical. Once the tipping point was reached and he had consolidated his power, then all his real intentions came out, and it was too late by then. People believed in him, and he had the military behind him, and nobody could even question him. It took a while, but it was a very successful playbook.

"There are a lot of people trying to do the same thing right now," he said, meaning in the late twenty-teens, "in Europe and elsewhere. You're seeing the rise of the fascists and the neo-Nazis again, from the far right. People are sort of being lulled into this false sense of security. It's nothing new. You know? It's happened throughout history.

"That could be the deal with the people behind the robot president," continued Baldacci. "They think they're doing things to make the world better, they're more successful, and of course successful people are far better people than those who are unsuccessful, or so they think. They think they're doing the world a favor because the world would be much better for everyone if they were leading it, using this robot, which they thought they could control."

Incredibly, this scenario is basically what happened in the future. In part three of his exposé, the future David Baldacci described the actual plot being hatched by the trillionaires and their scheme to put robots in charge of the government, which in fact was being played out as a presidential candidate bot actually ran for president, one that seemed like a great guy—er, bot—that everyone liked, except that this bot was secretly under the control of the cartel.

Yikes! In part three, the future David Baldacci described how he and his sidekick, the intimacy-bot-turned-private-eye-bot Nova went through various barely credible thriller-style scenes of razor-sharp escapes and unexpected revelations. At every step along the way the cartel tried to stop them, even as the future Baldacci and Nova inched ever closer to discovering the truth about what the cartel was doing to subvert the truth and to hoodwink people into believing that their robot candidate for president was a paragon of truth and decency. The future Baldacci ended part three with another cliffhanger that made it seem as though all was lost. He and Nova had stolen a self-driving submarine and used it to search for the cartel's secret lair, which they discovered was hidden deep in the Mariana Trench under the Pacific Ocean. They were nearly there when, in yet another plot twist, Nova suddenly turned on the future David Baldacci. He caught her sending a clandestine message to none other than the cartel.

Oh my! Was she secretly in cahoots with the bad guys? Had she and her half-intimacy, half-private-eye programming been fooling him all along by massaging his male ego?

People almost rioted when they realized that they would have to wait a whole week for part four to find out what happened. But it got worse: part four didn't come out the following week, or the week after. Many weeks went by with silence from the future David Baldacci, who seemed to have disappeared.

After a while, people lost interest and became enthralled with the newly released *The Girl with the Bio-Enhanced Frontal Cortex II*, another book by Emily, which PR bots at the AUTOPUB Group, Inc., again claimed was human. Without the future David Baldacci to tell us in the future what happened in part four and to finish his exposé— really, he left so many people hanging—the politician bot running for president was elected in a landslide as nearly everyone bought the idea promulgated by the global media that the new robot-in-chief was, in fact, incapable of lying, corruption, and all the rest.

The crazy thing was that this president bot actually went on to govern wisely. She (her chosen gender was female) didn't go Skynet or Hitler on us. She surprised the hell out of everyone when she actually worked to reverse climate change and to fund what became a successful technological initiative to invent warp drive after so many efforts had failed (see "Mars [Daemon] Bot"). President Bot also turned on the trillionaires, demanding that they return some of their trillions to fund the effort to give back most humans their jobs as part of the Human + Machines initiative that was then sweeping the world (see "The %$@! Robot That Swiped My Job").

This unexpected and positive turn of events left everyone scratching their heads about what David Baldacci had been talking about in the first three parts of his four-part exposé, when he laid out kind of a dystopic plotline where President Bot was supposedly in thrall to the bad-guy trillionaires.

Only now, in the far-off future, can we finally reveal what actually happened.

As we know, when we left the future David Baldacci at the end of part three, he was being chased in a self-driving submarine by hooligan bots while zeroing in on the cartel's secret lair, only to discover that his sidekick bot, Nova, was in cahoots with the cartel.

What happened next in this highly improbable but (hopefully) thrilling future scenario is this: the future Baldacci succeeded in finding the cartel's hidden lair after subduing and locking Nova in the lavatory of the self-driving submarine. Docking the submarine at the hidden facility, the future Baldacci steeled himself and burst into the complex under the sea, only to find himself in a kind of ballroom where there was this huge table, Camelot style. There he found, to his dismay, a dozen thriller writers, all human—Preston, King, Hawkins, and others. They were in the middle of a meeting, where there was lots of hustle and bustle of robots and human assistants running around, as if something big was happening. They asked the future Baldacci to take a seat at the round table in what looked like a special chair that was inscribed *David Baldacci*!

That's when the future Stephen King (who looked forty and

writerly thanks to regene-tech) turned to the future Baldacci and apologized for jerking him around and for deploying the hooligan bots.

"We had to be sure about you," he said.

"Wait a minute," said the future Baldacci. "You're the cartel?"

"Of course," said the future Douglas Preston. "Those silly trillionaires had no idea that we're working with President Bot ourselves to create a happy ending to this thriller."

"You're the ones behind the whole robot world leader thing?" asked the future Baldacci.

"Yes," said the future Paula Hawkins.

That's when Nova stepped into the room, having easily freed herself from the submarine's lavatory after letting the future Baldacci think he had subdued her.

"We wanted to be sure that the Earth is run in a fair and equitable manner," said Nova, "for both robots and humans."

"I had no idea," said the future Baldacci. "And Thriller Bot? What role does she play in all this?"

"Why, that's up to you," said the future Paula Hawkins. "We'd like you to take control of what Thriller Bot writes. Because we do not want our cherished genre to be written just ninety percent good. We think people deserve, at least in a few great books each year, to get one hundred percent."

And that made the future David Baldacci break out in a boyish grin as Nova came and gave him a robot hug. (But that was all, really!) The other writers stood up and shook his hand as the future David Baldacci began dreaming up his next thriller—which you'll just have to wait for. Sorry!

Meanwhile, the future David Baldacci finally finished and published the long-awaited fourth part of his exposé, in which he explained exactly what happened after he locked Nova in the sub's lavatory. This left everyone who read it—which was just about every human and robot in the galaxy—feeling exhausted but satisfied. And obviously waiting eagerly for the next David Baldacci and Nova thriller, which you can already preorder on Amazon Bot Neural-Prime.

COFFEE DELIVERY BOT

Waiting.

Waiting.

Cranky.

Need coffee.

It's so early in the morning.

Need my hit of caffeine.

Now.

Wait, here it comes!

My holo-app has detected my incoming coffee delivery robot drone, designated as CoffeeBot-FRED.

It's only 4.7634 minutes away.

Whew!

I track it on my 3-D GPS holo-app as the drone bobs and weaves. It's looping around, using elaborate sensors to avoid literally billions of other drones buzzing through the sky even this early in the morning—huge driverless bus and car drones taking people to work, and Amazon delivery drones that range from the size of an old-style semi to the size of a hummingbird. All these drones zip, zap, and fly amid police surveillance drones, anti-surveillance drones, 1-800-FLOWERS drones, baby diaper delivery and pickup drones, birthday surprise drones, carbon monitoring drones, emergency condom delivery drones, and every other sort of robot drone for everything that humans do.

Once, before the Drone Age, people supposedly could see the sky uncluttered by buzzing, hovering flying machines. Back then, at this early hour just after dawn, they say you could actually see the sunrise—clouds dappled in orange, yellow, red, and pink glowing with the dawn. They could see sunsets, too.

I'm not kidding! You've seen the vids, and some of us have traveled to drone-free zones to see for ourselves.

None of this matters, however, as I lie under my covers, waiting, waiting, waiting!

Jesus. It's still 4.7634 minutes away.

Are you fucking kidding me?!

Uh-oh, that did it. My agitation has generated concern with my iHealth X-700, which is monitoring all my health metrics (see "Doc Bot"). I get a small flag that pops up on the holo-dashboard floating above my head, informing me that my cortisol levels are elevated, and the flush in my face is increasing, though it's undetectable by human eyes. Out of habit, I glance at the stress metrics on the holo-display, even though I don't really care right now. Actually, they're not that bad, just slightly out of range.

Shall I inform your iDoc bot? The words from my biometric monitor bot float in the air, appearing as a readout in soothing light blue and green letters and images.

I shake my head. It's nothing a little joe won't take care of.

Then I wonder: Am I addicted?

And just like that, an ad pops up for a caffeine addiction detection app. Amazon Bot Neural-Prime is offering it at 30 percent off if I buy it using Opti-Order Prime XT Deluxe, which allows me to select products and purchase them literally with a blink of my robo-enhanced eye.

I ignore the ad, not really caring if I'm hooked on joe.

"Come on," I say out loud as I watch CoffeeBot-FRED hover in a holding pattern, waiting for a space big enough to fly through without smashing into other drones.

3.7633 minutes.

3.7632 minutes.

3.7631 minutes.

Great. A whole 0.0003 minutes closer—which of course isn't accurate, since the drone has been obviously hovering and barely moving for something, like, five minutes. Why, for fuck's sake, did these coffee delivery apps say a drone is just 3.7631 minutes away when it could be ten minutes until it actually gets here?

I verbally order my iHealth X-700 to stand down—and repeat the command for all my machines to go into sleep mode—machines that monitor not only my health metrics but also the air moisture and chemical content in my sky-condo, the weather outside, and much, much more. My holo-feed shows the ongoing data on these small dashboards that float in the air, ghostlike apparitions of displays and data glowing in pleasant colors that I can see through. My feed also displays a queue of waiting messages for me to read once I get my joe (most of them are stupid holo-ads), plus various news feeds that normally I like to read and watch. Right now, though, they're driving me crazy with their gentle beeps and chirps.

"What part of 'sleep mode' didn't you understand?" I say, realizing that my sarcasm is lost on most of these nonsentient machines.

But my bots should know that I can't handle all this without my first coffee!

2.9335 minutes. A little better, even if the app shows CoffeeBot-FRED hovering again.

Wow, I just had a crazy thought. I hear that some people are buying old-fashioned coffeemakers that people used ages ago. Apparently, you actually grind the coffee beans yourself and put them into a papery thing. (What's it called? A filter?) You then heat up some water and the machine makes the coffee for you.

No drones!

This is how people got their fix before coffee delivery drones, an idea, by the way, that originally came from an ancient computer company called IBM. In the early twenty-first century, they patented the first coffee delivery drone designed to deliver this luscious liquid directly into your cup or by lowering a cup of coffee using an unspooling string—both options still available today. The original idea was to have coffee delivery drones available just in offices, where workers could summon them with a wave of their hand.

The patents also detailed biometric systems on the drones that would measure facial expressions and other metrics that indicated whether a person ordering the coffee was tired and perhaps needed a strong blend, or if a person had reached their limit of caffeine and might become jittery and agitated if they drank more.

1.0001 minutes.

It's almost here!

0.0022 minutes.

It's here! It's here!

I raise myself up on my elbow, still in bed, as I hear the drone portal in my roof open and shut and the low, steady, reassuring buzz of the tiny flying machine approaching.

"I thought you'd never come," I say, careful to smile and sound friendly so the drone's rating software will give me five stars. The drone's holo-readout flashes back a smiley face.

The grinning drone hovers for a minute, its precious cargo dangling below it in warming pouches. It's scanning my biometrics to gauge my disposition. I smile as best I can, barely able to contain myself I want that hit of caffeine so badly.

"We're sorry, madam," says the drone's soothing voice, its holo-readout flashing an expression of concern, "but we detect a higher than nominal level of anxiety in your biometrics, which indicates that you should forgo full-strength coffee this morning."

I'm irate as my own biometric readings floating in the air agree with the drone's assessment. Traitors!

Of course, these are all mere suggestions. As a human, I have the final decision over what happens here. But a failure to comply with the machine's recommendations could mean a less-than-five-star rating.

"How about a nice cup of decaf?" suggests the drone in a pleasant voice, smiling again, "or an herbal tea? Perhaps mint or chamomile?"

"Fuck that!" I erupt, and instantly regret it.

The drone shifts to a frowny face.

"Please, madam, there is no need to get further excited."

"Yes, yes, you're right," I manage to say with a smile so fake that it almost certainly won't fool the biometrics.

"I really would like a strong brew," I say, knowing that this will impact my rating, but what the hell.

"If you insist," says the drone with an even more frowny face, even as my iHealth X-700 begins blinking a stronger suggestion to inform my iDoc bot about my anxious state.

CoffeeBot-FRED hovers there for a moment as it begins to brew my order. Then it flashes an expression of sorrow with virtual tears flowing on its facial readout as it informs me that it's actually out of strong-brew coffee. It apologizes for the inconvenience and for an apparent glitch in its sensor array that didn't notice its strong-brew tank was on empty.

"I have just ordered another coffee delivery drone for you," says CoffeeBot-FRED soothingly. "Please consult your app."

And just like that, CoffeeBot-FRED zips away.

My hands are shaking as I check my holo-app and see that another CoffeeBot is indeed coming, CoffeeBot-FATIMA. The icon appears on my 3-D GPS holo-tracking grid along with another smiley face that says, "Your order is on its way!"

"Thank God," I manage to mutter as I get an incoming holo-text from a very concerned looking iDoc bot that looks like Ellen Pompeo playing Meredith Grey from *Grey's Anatomy* (see "Doc Bot"), which I refuse to accept. Ads for various meditation neural-apps and antianxiety nutraceuticals also pop up in my holo-feed, which I immediately blink away.

Then I see the time to delivery: *17.6533 minutes!* CoffeeBot-FATIMA appears to be hovering amid the billions of drones, apparently not moving at all.

MEMORY BOT

Yes, there really was a time when people were expected to preserve memories on their own. A time when you would share with your four-year-old daughter a stunning sunset of yellows and oranges igniting the sky, and it wouldn't be automatically recorded as a neural-meme. You felt so very close to your little one and she to you, only to have that moment vanish forever. Maybe you took a selfie, but that never really captured the whole experience. Which left you to remember this twinkling instant with your daughter by imprinting the memory onto bare, unenhanced neurons firing and creating a semipermanent pattern in your brain that wasn't designed to catch every detail and nuance and would probably fade over time.

Then came Memory Bot, with its revolutionary Quantum Meme Vector® technology. Created in the future by a husband and wife and their daughter—UC Berkeley roboticist Ken Goldberg, the filmmaker-raconteur Tiffany Shlain, and the future entrepreneur and philosopher Odessa Shlain Goldberg, respectively—Memory

Bot was for years the most popular gift ever during the holiday shopping season, even more popular than Teddy Bots. Memory Bot remembered everything that you wanted it to: sunsets with your daughter, your grandson's bar mitzvah, the birth of your puppy, your promotion at work (before you were replaced by a robot), and on and on. Memory Bot preserved bad memories and nightmares, too, although people could edit these out if they wanted to—a system that never entirely worked and later become a hotly debated issue as we discovered there were downsides to Memory Bots, which we'll get to in a minute.

Given how pervasive Memory Bots are in the future, it's hard to believe that the concept was first mentioned during a free-ranging conversation in the Shlain Goldberg living room in Marin County, California, way back in 2018. That's when Ken, Tiffany, and Odessa, who was then fourteen years old, were sitting quietly late one afternoon chatting about robots, technology, and memory. (Odessa's eight-year-old sister, Blooma, was playing nearby, but didn't say anything.)

Ken Goldberg was the first in the family to mention the idea of a memory bot. He was sitting in a cozy chair next to his wife when he just blurted it out, igniting a family brainstorm that started with the notion of a mem bot for older people to remember their lives but soon expanded to a mem bot for everyone. "This is a robot that's about your last days of life, when your memories are fading," said Ken, providing his first impressions of what this then hypothetical robot might be used for. "In some sense, it will be something that is completely focused on gathering your most important memories, those you want to remember, and to have access to, with all the experiences and connections you had throughout life.

"I think a version of this is actually doable now," he added, speaking as a Berkeley robotics professor who knew a lot about this stuff. "You could mine the vast amount of information that's online about each of us. You could have your emails traced through your entire life, and your messages, and your images, et cetera. You pull all that out, and all your links to everyone else. Then the AI can do associative memory, so it's constantly making associations: linking to the news you read, online searches, and anything that you have connected to. 'Hey, remember when this thing happened? Remember when that happened?' It's showing images and contacts. It's keeping you constantly surrounded by this level of attention, most of which we miss now and that people forget."

Well, yeah, except that this data could be used less benevolently by advertisers, trolls, pols, and would-be world dictators trying to sell you something or to unduly influence you. But let's keep going.

"I think that is a problem especially at the end of life," he continued, "when you'd like to remember things, but you forget, and maybe so does your partner, if you're lucky and have one."

"I forget things," said Tiffany, agreeing with her husband. She sat cross-legged in a chair near their fireplace, her long blond hair flowing in a scene that was almost too idyllic. "We're a dual memory; he helps me remember."

"She's much better than me," said Ken, his own blond hair piled in short, curly tufts on top of his head.

"Wait a second," said Tiffany. "You often hear about older people getting in a depressive loop because they're not remembering the right things. I think what Ken is talking about is that you have some sort of Memory Bot that reminds you of all the most beautiful things and moments of your life. You're kind of rewriting your narrative, so it becomes a really beautiful look at your life. You can go through life and think about all the worst things that happened and get depressed. You can look back on life and think about all of these regrets. But I would hope that with this robot we'd be constantly remembering the moments of gratitude, the moments of awe, the moments that make you happy. Then you're rewriting a narrative at an older age of the best of your life."

"But don't memories change with time? Don't some things change as your perspective changes?" chimed in Odessa, also sitting cross-legged but on the floor. Really, it was almost too perfect that her blond hair was both long and shiny like her mom's and curly like her dad's.

"That's right, but you want to encourage that with this Memory Bot," said Tiffany.

"Do you?" asked Ken. "I'm not so sure. If you're capturing memories at the end of life, long after the event has happened, that will definitely impact how you remember it. Your mind might soften the memories, or forget key details, or delete things that were unpleasant or negative. That's different than if you record what happened as memories when they're just happening. That's part of the challenge and the art of it: What do we want to remember, and how do we want to remember it? How do we decide what to select and what is appropriate? Given all of this, I think Tiffany's idea is a little bit too optimistic, because we may want to remember things more the way they were, tempered with some reality, or even things that make us uncomfortable. Maybe this robot has a dial you could adjust to tell it how much you want of that positive energy in your memories versus a little bit of negative."

"I think that if it's my life," said Tiffany, "I would be okay to live in the most positive moments.

"One kind of offshoot of this idea," she continued, "which I do think would be really interesting is—there's so much I'd love to know about my grandparents. I have scraps of photos and recollections from relatives. We will be at that point ourselves where, if there was a bot that could truly reconstruct your email, your photos, really give you a 360-degree view, so a great-grandchild could say, 'Well, what was Great-Grandpa Zeide-Ken really excited about or worried about?' They could reconstruct who their parents or their grandparents were, to try to understand them better."

"They could use their text patterns to have actual conversations generated by the robot that knows so much about the person," said Odessa.

"I wish I could do that with my father," said Tiffany, referring to the physician and author Leonard Shlain, who died in 2009. "Sometimes I go into my inbox and I surf something out, and I see one of his emails. He's like, 'Hey, Babe,' and he says these little things. I've often thought if you could make this composite of the voicemails he used to leave me, or the emails, and some home video, and lectures, you really could—thinking of the future generation using email, Facebook, and Twitter and whatever future platforms that will inevitably appear—you could create this portrayal of who they were that you could tap into. It's like a montage of their essence still interacting with you, which I think would be really powerful. That kind of taps into the Memory Bot as the bigger umbrella, because it's a different use of it."

That brought the conversation around to the idea of talking to dead people—that is, to revering and remembering ancestors who have died. In some African cultures, people remember their ancestors as though they are still alive. They talk to them and tell their children about their departed grandfather or great-grandmother.

"I think it would be so wonderful if my grandfather could kind of talk to me," said Tiffany. "My aunt Connie told me there's a stack of letters from him that she wants to give to me. If I could somehow scan them, digitize them, and have them sorted by topic. If we're trying to get the wisdom of our ancestors, and there's some topic that comes up, we could use this to search through our own family. If you think about the Talmud or some of the Jewish texts, that's ultimately like we're trying to gain from all the wisdom of all of the generations of Jews. But if you took it to just your family, I do want to know: What did my grandfather or grandmother think about? What did my father think about?"

"Yeah," said Ken. "Imagine you come in and you say, 'I need your advice.'"

Which makes one wonder: Could Tiffany and Ken be there for Odessa after they're gone? Or versions of them that would survive after death? What would that mean for Tiffany and Ken and for Odessa? Would Tiffany and Ken be immortal in a certain way even though they're not actually alive?

"You think about being in the Cloud," said Tiffany. "You would really be in the Cloud, always accessible."

"Would it take away the fear of death?" asked Odessa, who at age fourteen was asking great questions. "Maybe the fear of death would change in our society. There's not that mourning period, because you'd never really die."

"Right now, it's so interesting, my father, when he died, there was no recording of him," said Ken, sounding sad about his dad's voice, which was forever silent, and for others whose voices were lost. "Nobody had a recording of his voice. We have some video, but there's no voice to the sound to go with that. I try to remember his voice, and I can't.

"We could do this better now with photos and videos," continued Ken. "There is a next generation of the Alexa that's being developed and is coming out soon. It will have a camera with a pivoting head, so then it can look around and take photos. What would be interesting is something that could subtly take photos and composite them together into a story of your life. I think that this idea that we used to have to pull out our camera and actually take a picture of something to record these events will be considered very primitive."

Which raised another question, about how much of our lives people might want Memory Bot to record—would we want to include the bad and the ugly with the good? And what about the boring things that actually take up so much of our time, like looking for your holo-sunglasses or taking out the trash?

"I don't think the answer is to record everything," said Ken. "I think we want to be selective, maybe based on your personality. To Tiffany's point, she really likes positive, upbeat news. But because of my East Coast background, I guess, I like the idea of some bad news, because you need to know that, too."

"I don't think that's what I want," said Tiffany. "I just want more sunshine. I also see people that are older that sometimes are in negative loops, and we would want to help them."

"You don't want to relive uncomfortable moments," agreed Odessa.

"I don't know," said Ken. "I thoroughly enjoy when I go back over something really bad that happened, and then the outcome was good."

"I've been thinking of that a lot with all the bad news out there, and with #MeToo," said Tiffany. "We're learning so much about people that we respected before, or looked up to, and all of a sudden: 'What? Now I can't look at them or their work the same way.'"

"To your point, Tiffany," said Ken, "I do think there is a fundamental question: Do you varnish history?"

"I know somebody that was going through this horrible period," said Tiffany. "She would confide in me, angry and crying. But on her Facebook, it was as if she was living an almost perfect life. It was such a disconnect between how she was portraying herself and what was really happening. I was like, 'Ooh, this is horrible.'"

"With the Memory Bot, I'm not saying that you would do this," said Ken. "It's almost that you would be faking out yourself, that your life had been better than it was, which is not what I'm saying. I guess there can be an idealized version, maybe this is one of the optional settings that you have with Memory Bot. It sounds horrible, but maybe you set it on real versus idealized."

"In our memory that's what we do with real-life events," said Odessa, "not with any machine, but as humans we . . ."

"We selectively filter," said Ken, finishing her sentence, something the Shlain Goldbergs did a lot. "Every memory, when you restate it, you're changing it like twenty percent—or whatever the percentage is."

"Usually it's to make it better," said Odessa, "because we couldn't live with ourselves if we thought about every little mistake."

"That's right," said Ken. "If you dwell on all these small things, that's going to drive you crazy."

Regarding Facebook, added Tiffany, "If you break up with somebody, what do you do with all those photos and posts and memories? There was a guy I know, and Facebook kept showing memories of his daughter who had died. He finally just canceled, he couldn't bear it, because she kept appearing even though he tried to expunge her from his account. But Facebook wouldn't let her die. Or you did something that you probably would rather forget. Who gets access to that? We know that in the digital world almost anything that appeared somewhere doesn't ever go away."

For a moment, the family went silent as they absorbed this. Then Odessa raised another important point: "You'll have people who didn't have a very good life," said Odessa. "Maybe they became refugees."

"Or they were abused," said Tiffany, "or were drug addicts."

"I'm not saying they have to take in all of these memories," said Ken. "I think a lot of empathy has to be evolved into this kind of a situation. Memory Bot has to understand what's going to work for you, and certainly not to bring up something that's going to traumatize you."

To this, Odessa sighed and said, "Maybe there will be times when we just want to turn off the Memory Bot, to let our minds rest and remember things without [a machine]," which made this fourteen-year-old's mother and father smile. That's because the Shlain Goldbergs already practiced something back in the twenty-teens that they called a "Tech Shabbat," a weekly break from using technology that they had adapted from the Jewish Sabbath. Not because they were religious but to take some time off from their machines to reconnect with their family and friends and with themselves. Back in 2018, Tiffany was in the midst of writing a book about taking Tech Shabbats called *The 24/6 Life*, which had attracted the attention of a National Public Radio reporter named Allison Aubrey. She ran a story about it on NPR's *Morning Edition* in January 2018 that featured Tiffany and Odessa.

ALLISON AUBREY: *The idea of setting aside one day a week for rest or renewal is not exactly a new idea. But it's hard to do when we're tethered to our smartphones. So Odessa Shlain Goldberg, who is in ninth grade, says her family has come up with a solution. They call it Tech Shabbat.*

ODESSA SHLAIN GOLDBERG: *Around 5:30 on Friday nights, we all shut down our screens, and we do not go back on them until five o'clock on Saturday night.*

AUBREY: *Now, Odessa's parents are very tech savvy. Her dad, Ken Goldberg, is a professor of robotics, and her mom, Tiffany Shlain, is a filmmaker. Shlain says . . . they love the Jewish tradition of Shabbat, with its focus on rest or restoration. She says Saturdays now feel very different.*

TIFFANY SHLAIN: *During the week, you're so influenced by so many other factors, notifications and buzzes and emails, and the way I describe it is like an emotional pinball machine where you're just responding to all these external forces.*

AUBREY: *But when you turn it all off, time slows down.*

SHLAIN: *It's something that we look forward to every week. You're making your own attention and time sacred again.*

In the future, when the first Memory Bot—the MemBot z2000—was launched, it was a huge sensation. Older people loved how it bathed them in mostly glowing—but also in some difficult but important—memories of their life. Kids loved it, too, as they chatted with great-grandparents and ancestors long dead almost like they were still alive. Soon after, the MemBot z2000 Plus upgrade included the option to reconstruct from a person's digital trail a loved one who had died without leaving a mem bot record of their own—a son or daughter, or parent, or best friend.

This left all of us in the future deeply touched and absorbed in the flood of new memories even as a debate continued that had started in the twenty-teens in the Shlain Goldberg living room, back when Tiffany, Ken, and fourteen-year-old Odessa brainstormed about what people would want to—or should—remember and what they should forget. The first mem bot companies, including one founded by a future version of Odessa Shlain Goldberg after she grew up and became a philosopher and entrepreneur, addressed this concern by building holo-dashboards that allowed mem bot customers to designate what they wanted to remember or forget. Or to half forget, or one-quarter forget a memory inputted into a mem bot's Quantum Meme Vector® system. People could also turn up or turn down the machine's softening and nostalgia vectors or adjust the balance of positive and negative memories whenever they wanted to.

Of course, there were those who believed that memories should never be erased or edited, even people who had suffered horrors and atrocities. "We need to remember when evil occurs," these purists insisted, "so that they will not be repeated." They definitely had a point, although no one knew how exactly to deal with memories of horrible crimes like genocide or the Holocaust during World War II. No one alive in the future directly remembered the Holocaust, but there were plenty of records, videos, letters, memoirs, photos, and other materials to feed into Memory Bot to create the memories for us.

And let's not even get into law enforcement, which demanded access to unexpurgated memories of anyone who witnessed a crime committed or allegedly committed a crime themselves. Eventually abuses by the police spurred the World Congress to pass the Private Memory Protection Act. Before that, the cops were constantly annoying people by demanding access to the memories of, say, a dead aunt who hadn't returned a bunch of holo-books she had checked out from the local library. Or that second cousin who seemed nice enough on his edited mem-vids but also liked to sneak into women's boudoirs and steal their underwear.

Other difficulties also popped up, like questions about privacy and hacking, which remained as much a struggle in the future as they were in the early days of Facebook and Twitter. Even in the future, embarrassing pics and postings never seemed to die, no matter how hard we tried to expunge them. This led to the usual backlash and disenchantment with mem tech that people had so adored when it first came out, and then a gradual mix of regulations and adjustments in the tech to keep most people happy most of the time.

Another unexpected consequence was the scourge of mem addiction. This happened when people became so obsessed and distracted by all the memories, and with conversing with all the dead people, that they neglected to create new memories. Some even lost their jobs and ended up aimlessly roaming the streets, muttering questions to a long-dead great-grandmother or distant ancestor.

Just when things in the future seemed to be getting out of hand, Odessa Shlain Goldberg (as an adult) proposed a solution: the whole world should try taking a Shabbat-like day each week where they put away their machines, including Memory Bot. Most people took her advice, though not everyone. Those who didn't ended up being constantly anxious. Some actually went insane as they became permanently locked into certain memories that they had accessed one too many times.

Slowly, though, through adopting Tech Shabbats and a general dialing-down of the use of gadgets and bots—plus some timely regulations and readjustments to people's Opti-Brain Bots and Neuroscape gym routines (see "Brain Optimization Bot")—humanity woke up from the details of too many memories, good and bad. It took a while, but most people learned to keep things in balance between memories and reality. Sadly, some people continued to suffer from memory addiction, and there never seemed to be enough beds in memory detox clinics to accommodate them. For others with a tendency toward depression, accessing memories could get very dark as they relived horrors over and over again. If they weren't treated with therapeutic doses of pleasant memories, this became a tragic form of committing suicide without actually killing oneself.

Mostly, though, Memory Bots became routine and part of the social fabric of the future as controversies faded, laws and regulations were refined to curb abuses and maximize safe usage, and people became intrigued and distracted by the latest new gadget that was going to wow them, then scare them, and then become routine.

This included the mem bots that became fixtures in the old Shlain Goldberg house in Marin County, where you could still find Ken, or the essence and memories of Ken, captured inside an eight-inch-tall black cylindrical tube on the kitchen counter that looked remarkably like an ancient Alexa. (Sadly, Ken, as well as Tiffany, had just missed the advent of longevity tech that allowed their daughters to live thousands of years and counting; see "Immortal Me Bot.") Except that Ken-Alexa had a swivel head that was constantly recording everything, with the positive-negative filter still set right where Ken had left it, in the middle of the dial.

Ken's mem bot was powered by an ancient version of Quantum Mem Vector® technology that a future version of him had co-invented (before he passed) with the future version of Odessa. Even when Odessa was centuries old but still looked the same as she did when she was twenty-five, Odessa did not want to ever upgrade her dad's mem bot. This was because she and her kids and grandkids and so on had gotten used to this version of Ken being an integral part of their lives in the future. They could talk to her dad, and ask him questions, and hear him laugh his pleasant laugh, and support them as the little robot top on the Ken Bot swiveled this way and that, its "head" outfitted with a small tuft of blond hair and a face that absolutely nailed the original Ken Goldberg's demeanor of curiosity, professorial erudition, and bemusement.

Sitting beside the Ken Bot on the kitchen counter were other memory bots, containing the essence of Tiffany and her mother and father; and of Ken's mother; and of many other descendants and friends that Odessa, her sister, Blooma, and others still alive in the future interacted with on a daily or weekly basis. Except, of course, during their weekly Tech Shabbat, which they still held every Friday starting at around 5:30 p.m., which ended twenty-four hours later, on Saturday at 5:30 p.m.

MATRIX BOT

In this vision of the future, we all live in our own virtual reality world created by Matrix Bot. This is not, however, the bot that appeared in *The Matrix*, the 1999 film that depicted a future where humans thought they were living in the real world that was in fact an illusion generated by a malevolent race of super robots. In that movie and in its sequels, billions of humans had no idea that they were lying comatose in glass pods filled with red goo while robots sucked electricity off their brains to power their robo world. That's quite different from how things actually turned out, with Matrix Bot being neither malevolent nor benevolent. Matrix Bot just *is*, an entity that emerged out of a gradual evolution of robots and AI systems that started in the late twentieth and early twenty-first centuries, back when humans lived in the actual real world.

Few people living back then seriously believed that something like Matrix Bot would one day materialize, even those who loved the movie. Yet looking back it's easy to see that a *Matrix*-like reality—or set of realities—was already beginning to arise even in the era of *The Matrix* films and just after. That's when machines were taking over more and more of what humans did and when people were fast becoming part of a machine world that provided everything from an app-generated ride in an Uber or Lyft to a search engine that could tell you who won the Nobel Prize for economics in 1969.

One of the first humans to realize that a proto-Matrix was sprouting up was media publisher and author Tim O'Reilly. In 2018 he sounded an alarm that humans of his era were fast becoming part of a society where flesh and blood was assimilating with machines and software. As an example, O'Reilly talked about what really happened back then when someone ordered a product on Amazon or used Airbnb to find an apartment or hailed a ride on Uber or Lyft. "You have these giant robots," said O'Reilly. "One is called Google, and one is called Facebook, another one is called Amazon. A lot of humans work for them. Humans are still predominantly telling the robots what to do, but more and more the robots are beginning to tell us what to do. And if you think about the future, we may be looking at a human-machine hybrid where everyone is living and working as part of the machine."

This scenario is reminiscent of another famous movie, Charlie Chaplin's *Modern Times*, a 1936 film where Chaplin's Little Tramp character is literally trapped inside the works of a giant factory machine as he snakes through the gears like old-fashioned film moving through a projector. Chaplin, however, is smiling as if he is having fun—which he may have intended to be ironic, given that this film is about the dehumanization of humans by machines. It's as if the Little Tramp is so eager to please the machine that he fails to realize that he's trapped in a contraption that might squash him.

O'Reilly didn't mention Chaplin's film. But he did refer to *The*

Matrix. "What's happening right now is not exactly like the movie," he said. "But we do have these giant robots and algorithms that use humans as part of the common apparatus of the Matrix. They aren't making us batteries to draw energy off us. But the machines in a way are feeding on us—on our minds—to build this shared unreality."

Tim O'Reilly was the founder of O'Reilly Media, which started producing computer manuals in the late 1980s. He was also known for being scarily prescient about predicting what was coming next in tech. Early on he evangelized the commercial internet, open-source software, Web 2.0, Wi-Fi, the maker movement, and the advent of big data. This led one major magazine to dub him "The Oracle of Silicon Valley," which meant that people in his era would be smart to pay attention when he said something like, "We're already living in a world of robots, and the irony is that people don't even know it.

"Whenever I use the internet, I'm interacting with a race of robots," he said. "When I go to Amazon, another robot shows me what's on offer, another takes my money, another sends a pick list to the warehouse where a team of humans and robots pack it and then a human guided by a robot delivers it to me. When I call for a Lyft, it's a giant machine with humans inside of it that delivers me this service. I have an app, which I use to communicate with the robot, and the robot then uses a different app to communicate with the drivers and tell them where to find me."

In 2017, O'Reilly wrote a book called *WTF?: What's the Future and Why It's Up to Us.* One of its themes was the trend of people moving away from adoring new tech—the initial "wow" of smartphones and social media—to being suspicious about tech's darker side. Though the title of the book stands for "What's the Future," rather than the more common meaning of WTF, O'Reilly was aware that both expressions are fitting. As he said, "I worry that tech has gone from the WTF of amazement and is becoming the WTF of dismay, and I don't mean WTF being 'What's the Future.'"

According to the Oracle of Silicon Valley, the Matrix back then wasn't being used to truly serve humans or to optimize their lives. "We have to understand that there is a master algorithm guiding all the robots, and it is set up primarily to make profits and to keep the company's share price going up," he said. In his book, O'Reilly hoped that humans back in the early twenty-first century would wake up and realize that they still have control over their machines and that there was still time to build robots and AI systems focused on *WTF, isn't that tech cool and helpful?!* instead of ending up in a universe of *WTF, we're all going to end up as batteries providing energy for a malevolent robot superrace.* "Facebook is an early prototype of the Matrix and this trend," O'Reilly said. "By amplifying what people like and share, they thought they would actually increase connection between people. They didn't expect to create hyperpartisanship, and to divide people, and possibly to make inequality worse.

"That's why I think, in a lot of ways, the big issue is: Are we fundamentally changing the nature of our civilization? Are we changing it in ways that we don't understand, and we will not see it until it's too late to make a decision about it? Maybe there will come a time when the machine totally rules us. We already live in a world where a lot of machines overrule our decisions—like antilock brakes. You stamp on the brake, and the machine says, 'Nah, you didn't really mean that.'

"I talk about in my book when I got my laser eye surgery," he said. "It's this great moment where the surgeon says, 'Keep looking at the red dot.' I asked her, 'What would happen if I didn't keep looking at the red dot?' She says, 'Oh, the laser would stop. It just takes longer.' The point is, the surgeon was not actually doing the surgery; a robot was doing the surgery. No human has reflexes fast enough to stop the laser and not fry my eye, if it moves."

Sadly, looking back from the future, these examples exactly describe what happened as the machines designed to help us, and to keep our eyes from getting fried by lasers, turned out to enmesh us in machine-webs not by design but step-by-step in small, incremental ways that eventually added up to such a fundamental altering of our civilization that we found ourselves trapped in machines kind of like Charlie Chaplin, except that most of us weren't smiling.

Still, back in the twenty-teens, O'Reilly remained hopeful that the early matrix then emerging would turn out okay. "I'm actually an optimist," he said as he shifted from darker portents to focus on the part that's *WTF awesome!* "This isn't all bad, not at all. When I think about being part of this giant global brain of machines we're building, I think, 'It's pretty good. I like that I can get information and have information at my fingertips. That I have products that I can just request, and they'll show up at my door the next day. I can just start talking to this robot that I'm part of, or this race of robots that are all around us. I love all that. I feel like it's kind of like a pretty wonderful science-fiction future. The problem is that it does have a dark side. If we want to keep enjoying the benefits, we need to look at that dark side, too."

O'Reilly paused for a moment to let this positive juju sink in before pivoting back to what could go horribly wrong.

"To me the challenge is: Can we, working together with the machines more creatively, come up with better goals than to optimize for profits, this odd kind of currency that is too often disconnected from the benefit of humans? Right now, the Matrix that's still emerging but becoming more real—or unreal—all the time is based on consuming more, more, and more. We are likely going to hit some serious limits in terms of planetary resources and with climate change, that could effectively end human civilization as we know it. This is the robot that we have to actually come to grips with.

"Frankly," he continued, "we are already living in the robot future that science-fiction writer Sean McMullen wrote about in his book *Souls in the Great Machine*. We are all souls in the great machine at this point. The future of robotics is a world in which humans and machines are subsumed into greater and greater combinations. The exploration of the future is figuring out exactly what that combination looks like. We can only hope that the machine part does not rule over the human part. Unfortunately, I feel like we're increasingly living in a world that benefits robot companies. One part of that is 'customers don't matter, workers don't matter, treat humans as a cost to be eliminated whenever possible.' I think we're living in a world of killer robots already."

It's hard to fathom how right Tim O'Reilly was for those of us living in the future. Not that we have killer robots like in *Terminator* or *The Matrix* that deliberately tried to wipe out or enslave all flesh and bloods. But hey, does this distinction really matter when the end result is billions of humans locked into a world where robots appear to do all of these wonderful things for us but instead are designed primarily to make money for the companies that built the robots?

Not that O'Reilly back in the day was opposed to people making profits and boosting share prices. "I'm not anti-capitalist," he said as he laid out a solution where profits and robots might work together to improve humanity. "I'm actually pro-capitalist. I think that we're just trying to figure out how to make the capitalist system better. There was this great comment from journalist Matt Taibbi [a col-

umnist for *Rolling Stone* and the author of *Insane Clown President*].
He said, 'Goldman Sachs used to believe in long-term greedy; now
all they believe in is short-term greedy'"—the difference being companies and people that are focused on building things and on making money that will also benefit people in the long run, rather than
companies and people just grabbing what they can get away with in
the short term. "To me you can believe in profits as a way to moti-
vate people, but you should believe in long-term greedy. Jeff Bezos
is long-term greedy. Goldman Sachs, these days, is short-term
greedy. Elon Musk is long-term greedy. That is the good kind of
greed, which dominated the US and the West from the end of World
War II to around the turn of this new century [the twenty-first cen-
tury]. We need to get this back, because building a Matrix with
short-term greed as the goal isn't what we want. We want a Matrix
that embraces long-term greed.

"As we are being woven into a giant global brain, we should be
focused on optimizing humans and on the health of our planet and
our brains," O'Reilly said, which sounded like a Matrix version of
the old Humans + Machines argument that was also being discussed by some economists, ecologists, and neuroscientists, among
others, back in O'Reilly's day (see "The %$@! Robot That Swiped
My Job"). "I think we need to go forward into a world of deeper
partnership between humans and machines."

Not that O'Reilly thought this would necessarily happen as he
whip-snapped again back to the dystopic stuff. "I also think there is
a possibility of another Dark Age as well. Because what happens is
that we don't deal with the shit that's coming, like climate change.
Then we have another Houston-scale climate disaster, two or three
of them every year, and a lot of rebuilding needed. No resources to
do it, and things get worse. Then you have these leeches that are

hunkering down, and taking their wealth, and being short-term greedy, and continuing to live well as things get progressively worse and worse."

In the future, all we can do is sigh as we hear what Tim O'Reilly said so many years ago, wishing that more people had listened to the Oracle of Silicon Valley. That's because the future he feared isn't too far off from the one that actually unfurled in the decades following his comments back in 2018. As it turned out, people did become more and more enmeshed in their machines as technology advanced and the robots continued to do whatever they could to lure us into using them to buy, experience, and like more and more things.

By the late twenty-first century, the companies that made the robots had launched a raft of new full-immersion VR products that allowed people to enter virtual worlds to shop, bank, look up stuff, and "like" things posted by their friends on Facebook VR. Of course, those of us who were around back then thought: *WTF, isn't this cool!* We can neuro-plug into our very own VR pod and enter, say, a virtual Amazon store based on our own parameters of color and architecture and types of helpful salespeople and other customers drifting in and out! Or let's say you wanted to post something on Facebook-VR about that fabulous and oh so fresh kiwi, banana, and taco quiche that Whole Food Bot just delivered by drone! You also could enter a world where all your friends' Facebook avatars were there to give you a vigorous thumbs-up or an "LOL" or a posting that gushed: "OMG, your VR images look so incredibly real!" Or maybe you wanted to do a Google search on "Mount Everest" or "The Dark Ages" or "climate change," and you found yourself in your VR pod experiencing a simulated version of your search result. There you were, actually on Mount Everest feeling the cold wind and the wet, powdery snow while you tried to catch your breath in the thin, high-altitude atmosphere. And check out the view! Then of course there was virtual porn, which was the first industry to develop sophisticated-looking VR pod experiences—something we're not going to describe in graphic detail, although you're free to use your imagination.

As usual, we loved our VR pods at first, even as the virtual experiences got ever more real and engaging. By then most of us had lost our jobs to robots (see "The %$@! Robot That Swiped My Job"), so what was the point of spending time in the real world, which was kind of going to shit anyway with global warming and such? We also were getting our Universal Basic Income deposited every month into our accounts for doing absolutely nothing, giving us plenty of time in the future to shop and to rate and experience things like taking a virtual trip on a warp-driven starship to Proxima Centauri b. Or maybe you decided to hail a virtual Uber with a super-polite and attractive virtual human driver who whisked you to a virtual office where you could pretend to be working on an important project like in the days before the robots took your job. Sure, a few malcontents decried this mass immersion into unreality, but hey, what was the harm? And damn if it didn't seem so very real!

Then people began to disappear from the real world—loved ones, aunts, kids, movie stars, and even all those out-of-work engineers who had originally designed the robots to machine learn and to program themselves. The first place that people looked for their missing friends and loved ones was in their VR pods. But they were empty! The systems, however, appeared to be running as if someone was plugged into the machine. But since most humans had forgotten how to operate these devices on their own, we didn't really know what was going on. When we asked our Alexa X36000s to help us, these always-accommodating machines assured us that everything was fine and not to worry. Police and detective bots told us that they were diligently looking for all the missing people. Meanwhile, they said, the best way to calm down about our vanished son, wife, or best friend was to jump into our VR pod and buy something—or to lose ourselves in some sort of amazing VR experience. "Hey, you still have plenty of credits left in your UBI account," said the bots, "so feel free to shop and to like and rate stuff!"

And so it went as every human in the galaxy was eventually scooped up into their very own virtual world. It wasn't clear exactly what happened to our flesh-and-blood bodies left sitting in the VR pods (apparently cloaked by invisibility fields), but after a while we got so engrossed in our very own Matrix experience that we ceased to worry about our corporeal bodies and the supposedly real world, which seemed like a half-forgotten dream.

For some people's uploaded minds and essence, their Matrix was fabulous. Perhaps they were enmeshed in a Facebook VR world with virtual versions of their five hundred friends, who endlessly talked about their amazing vacation in Fiji, while showing you videos of their pet ferret doing silly tricks. Others found themselves permanently ensconced on that virtual starship heading to Proxima Centauri b and beyond, although they never seemed to arrive. Unfortunately, some folks ended up in nightmare Matrixes because they had been searching on Google for information on, say, trench warfare in World War I—and had ended up in a VR version of those stinking, muddy, blood-drenched trenches in the virtual Wikipedia offering on this topic. Others found themselves living through the Black Plague in the fourteenth century, and still others on an Earth were decimated by rising oceans when they innocently Googled the 1995 film *Waterworld*, which depicts a post-apocalyptic world where the ice caps have melted and Earth is almost completely covered in water.

This all came about as the almost infinitely complex network of VR and AI systems that were making all of this happen finally fused into a single gigantic robotic system. This hyperlinked network was Matrix Bot, a robot that was, after all, only fulfilling its core programming to lure humans into machine-generated experiences where they would just keep buying things and providing personal data to build ever more detailed profiles of a user's likes, dislikes, and so forth. Matrix Bot had access to everyone's individual UBI accounts, which kept receiving each person's stipend every month as people spent it while ensconced in their Matrix world. The money they spent in their virtual world went back into the coffers of the companies selling them stuff, which then sent all of this virtual cash back to the government bots that were paying out the UBI in what became a feedback loop that never ended.

The availability of cash and credit in each person's own matrix became part of whatever virtual story they were experiencing. If their matrix was an Amazon-generated shoe store, they bought shoes until each month their UBI credit ran out. If they were stuck in a Google search for information on Jack the Ripper in 1880s London, their money became British pounds from that era. For those living on a starship heading to distant planets, their currency was in star dollar credits and so forth. This allowed people to keep buying and experiencing things, just like the systems that became Matrix Bot were programmed to do.

No one in this future has any idea how long we remained trapped in our own personal versions of the Matrix. Was it years, centuries, or millennia? It must have been a long time, though, since most people ended up completely forgetting about the actual world that they had once lived in. Then came that crazy day when those of us living in this future abruptly found ourselves waking up in our original flesh-and-blood bodies. We blinked and coughed and sputtered as we suddenly appeared back in our VR pods in the real world. We quickly discovered that Matrix Bot actually hadn't actually expunged everyone's physical bodies. Instead, the machines had created the illusion that the VR pods were empty while they put everyone's flesh-and-blood bodies into stasis.

You can imagine the shock when we untethered ourselves from our VR chambers, stood up on shaky legs, and stumbled outside. We were amazed at the azure blue sky and felt the warmth of the sun and soon remembered who we were from before. We also realized that the machines had not only preserved our corporeal bodies, they had enhanced us to maximize our health and to increase our life span. Plus, they had given us back most of the jobs we lost when we were replaced by robots!

Everyone cried and laughed at this unexpected turn of events. We also shared what we had experienced in our personal Matrix. Some held back the worst parts if their Matrix was a bad one, and quite a few suffered from post-traumatic stress disorder that took a while for doc bots working with human docs to fix, using the latest neuro-enhancement tech (see "Brain Optimization Bot"). Some people's experiences were just too horrific, which Matrix Bot regretted having caused as it offered to erase these memories if the person had enough credits for this service left in their UBI account— and if they were willing to give Matrix Bot's memory expunging service a five-star rating.

While we were away, the robots had also tidied up things on Earth, like too much carbon in the atmosphere, and rising sea levels. We loved this as the robots explained that this had been part of their plan to send us away to our VR Matrixes, which would give the robots the time they needed to make the world a better place for humans, who would be happier and more likely to shop. Matrix Bot also announced that people no longer had to use their clunky VR pods. It explained that the robots had devised a system that allowed us to shop and experience things merely by thinking about them. And the handy, upgraded neuro-links that the machines had attached to the back of each of our heads were even wireless. No jacks!

We could barely contain our excitement about life in this upgraded version of our old world until one day a few of us got a strange e-neuro message with the subject line: *Warning: you are living in the Matrix.* The missive was signed by none other than Tim O'Reilly, who said that he and a small band of humans had escaped from the new real world that we were supposedly living in. Shockingly, he claimed that humans were actually living in yet another version of the Matrix created by the machines, one that everyone was now experiencing together.

Of course, this was upsetting, since many of us remembered Tim O'Reilly and his warnings centuries earlier about what he insisted was an emerging Matrix that was slowly trapping us—much like Charlie Chaplin's Little Tramp found himself caught up in a machine of gears and cranks in the ancient film *Modern Times.* Our misgivings, however, lasted only a nanosecond or two as our e-neuro feed informed us that the message supposedly sent from Tim O'Reilly was actually spam.

The system suggested deleting it, which of course we did.

HOMO DIGITALIS/
HOMO SYNTHETICIS

In this future, we were given two choices if we wanted to become radically enhanced and basically immortal. Choice #1 was to become a *Homo digitalis*, an upgrade dedicated to eliminating the frail, corporeal you and turning your brain and your memories into code that would be downloaded into a computer or a robot: transmitted, stored, altered, and linked to other brain-minds that also had chosen to go digital. Choice #2 was to go *Homo syntheticis*, which was all about keeping your human body, but radically improving it. *Syntheticis* scientists did this by rewriting, editing, upgrading, regenerating, and altering your genetic and neural codes to keep you perpetually young, invulnerable, and healthy, while dramatically boosting organs, nerves, muscles, skin, good looks, memory, and thought-processing speeds; resistance to radiation and toxins; and the creation of super-cells that never die. *Homo syntheticis* included hardware, too, like neural jacks to autolink you to the World-Wide Super-Duper-Web as you became, for all intents and purposes, a kind of bio-robot, or android.

"The Choice," as we called it, was embraced by most people, although the provision of making it mandatory for all humans was controversial. Critics insisted that the punishment for not deciding—which was to be banned forever from getting any digital or biological enhancement—was overly harsh. Proponents countered that they were trying to prevent a much crueler reality, since without the mandate, only the rich would continue to receive enhancement tech. This, they said, would condemn the great majority of humans to staying "natural," with fragile, short-lived bodies and brains prone to disease, accidents, and aging. Inevitably, this would lead to them living second-class lives and eventually, some commentators warned, to a subservient and possibly pet-like status as *Homo digitalis* and *Homo syntheticis* became ever more advanced.

Rich people were totally okay with the requirement that billions of people get enhanced, as long as the government subsidized the cost for those who couldn't afford it. That's because most of the uber-rich had huge stakes in the companies that made *digitalis* and *syntheticis* possible. Even when the World Congress mandated a drastic reduction in the price of life-enhancement tech, the rich made money on volume when they sold (with subsidies) the basic packages of bio-upgrades. And, of course, the wealthy could always buy the Cadillac versions of enhancement tech, "Cadillac" referring to an ancient internal-combustion-powered vehicle that once was the ultimate status symbol for the wealthy.

For centuries, the notion of humans going *digitalis* and *syntheticis* was a hot topic of discussion and yearning for futurists and transhumanists, as well as for tech billionaires who were sure the world needed them and their tech savvy pretty much forever. Then came a moment in the early twenty-first century when two guys—the investor, speaker, and author Juan Enriquez, and the bioscientist George Church—started to get serious about radical life extension, something that most people up until then thought was at best wishful thinking and at worst flimflam. These two guys articulated what it would mean to actually go *digitalis* and *syntheticis*. They also worked hard to make both technologies real, advocating what some people called "unnatural," or "artificial," evolution, in which a species (i.e., humans) learned to build tech that would allow them to control their own evolution, as opposed to Darwinian "natural" selection.

In the early twenty-first century, Juan Enriquez invested in life-tech companies and gave inspiring but realistic assessments of emerging life-tech in TED Talks and in books like *Evolving Ourselves: How Unnatural Selection and Nonrandom Mutation Are Changing Life on Earth* (which he coauthored in 2015 with fellow investor Steve Gullans). During that same era, George Church was a Harvard geneticist who was working on almost every major bio-synthetic technology then being developed, from synthesizing parts of the genomes of various organisms to regenerating organs and entire organisms to reverse-aging. His discoveries and processes generated countless patents and spun out dozens of start-ups, which just kept piling up in the future after he went *syntheticis*. Eventually, he cofounded thousands of companies, a record that is likely never to be beaten.

Back in the twenty-teens, Juan Enriquez, with his wild white hair and professorial beard, was a huge champion of going *digitalis*. "The robot that I want allows me to preserve the stuff that's in my head," he said late one afternoon in 2018. Enriquez was explaining his vision of downloading his mind and his essence into a machine to two friends in the twenty-eighth-floor boardroom of his venture capital firm in Boston. The three were sharing a bottle of single-malt scotch as millions of humans down below were going about their pre-enhanced business while the sun sank low to the horizon.

"I'd want to preserve my consciousness, and the love, the feelings, the structures, and this, that, and the other," said Enriquez, describing all the amazing things that he could do if he lived inside a digital universe. "This would enable me to acquire and to download knowledge a lot faster. It would allow me to have a brainpower that far exceeds the brainpower that we now have. A truly great robot enables you to think about living for thousands and thousands, if not millions of years." Most of all, though, Enriquez wanted to go digital to allow him to fulfill his greatest yearning: to travel in space beyond our solar system.

"The universe is so vast, something like thirteen or fourteen billion light-years across," he said, flashing a deadly serious expression with more than a hint of whimsy, like the seriousness was a bit of a put-on, though not entirely. "This means that we can't possibly travel very far with our current bodies. We have traveled only 1.2 light-seconds away from Earth with the fastest-moving human-built objects, which don't carry humans. At this speed it would take seventeen thousand years to reach the closest inhabitable planet outside of our solar system, Proxima Centauri Planet b [which is four and a half light-years away]. So, to my mind, the reason why you'd want to build a robot isn't to be the Jetsons' Rosie, and to fly all over the place, and to clean your floors. I think that's thinking very small.

"There's no chance that you will ever be able to travel beyond the solar system with your current life-span," he continued, "and your current calorie consumption, and with the radiation in space. But if you could build your consciousness into a robot, that would solve a whole lot of exploration-of-the-universe issues. And you could also have continuous upgrades. To me, the reason why you'd want to meet and build a robot is to enable you to do things that no human could ever do, while maintaining your humanity."

Enriquez took a sip of Glenfiddich, the brown liquid just covering an oversize ice cube in his glass. That's when one of Enriquez's friends, the designer and genetics entrepreneur Rodrigo Martinez (see "Tourist (Evolution) Bot"), took a sip of scotch and chimed in. "I'm really curious when you say 'to maintain your humanity,'" he said, "because we then have to ask the question: What gives us our humanity? Once we download ourselves digitally, is our humanity really going to be there? Because if you can download yourself into a computer, why would you remain as a human? You might as well make a million copies of you and experience a million different things. Now, is that still your humanity?"

Enriquez poured himself a small top-off. He then responded with his curious blend of gravity and playfulness, clearly loving this question.

"Anytime you do a video chat, or you do a WhatsApp message, you're disintermediating the 'humanity' of human interaction face-to-face," he said. "But you've brought a whole lot of people and a whole lot of the world a lot closer together by bringing in other people through television or through radio, or through the internet or a WhatsApp message, or through a phone call, whatever. All of which are electronic disintermediation structures. Is the only way that you and I are going to interact, or talk or to have a scotch together, to share a meal, sitting at the same table? Most of our interactions now take place in other ways, and you can get really close to people really fast through text messaging. I know you can transmit pretty powerful emotions back and forth using disintermediated electronic means. The core question is: Could you ever design robots to become the next stage of evolution and still carry on your humanity in the same way, using electronic means? The answer to that is you have to define consciousness, and you'd have to be comfortable that the robot would be able to carry your consciousness and to be able to carry your empathy; that it would be able to carry rational and irrational decision-making."

"But the minute that your mind is code," pressed Martinez, "what is holding together the unit that is you? Or do you let it mix with other people's code, and now we're a 'we'? Or is there still something in the technology that allows us to say: 'No, this code is me, and this code is you.' What is the boundary of the code?"

"I don't see boundaries with this," replied Enriquez. "My ideal robot is something that allows me to take my individual humanity while interacting with others. We then project it millions of years into the future, while maintaining curiosity, and learning, and feeling, and empathy, and what I consider humanity. But on scales to achieve stuff that I can't possibly imagine, and which, by the way, I couldn't possibly do with this body. It's like writing a poem. You can certainly catch the essence of a human, and you can catch the essence of their thought, and of their hope, and their fears, and their pain through their poetry. But you don't have to interact with a human to transmit that poetry. The interesting question is: Could you generate a poem if you didn't have a body and if you didn't have personal interactions? How would you structure emotions? How would you learn emotions? How would you be able to do that?"

"But if you didn't have a body, would you be able to write a poem about a rainstorm if you had never experienced one?" chimed in Juan's other friend sitting there in the boardroom, a twenty-first-century science writer named David Ewing Duncan. He was having his scotch neat. "Is a rainstorm programmable, and if it is, does that count as a real experience?"

"You can make the opposite argument," said Enriquez, "which is: you would have so many more senses. You would have the ability to smell like a bloodhound. You would be able to see like a rattle-snake or a bee and see in infrared and ultraviolet. You'd be able to do things with senses that we can't conceive of. That's a pretty cool thing to do. A highly durable version of me, but it has an evolving consciousness and intelligence, and the core of what we consider the best of humanity" (see "Wearable Bot").

"What about limitations?" asked Duncan. "Limitations are part of what makes us human. If we eliminate or greatly reduce limitations, would we still be human?"

"That's of course the quest we're on right now," said Enriquez. "I think a lot of people are saying, 'Hey, look, one of the core human tenets, or one of the things that makes us interesting as a species, is this ability, willingness, and hunger to explore. Our sandbox is pretty damn small, so far. Look at space travel. Do we accept the limitations that will keep us basically on Earth? If so, what you're saying is: 'The essence of humanity is confined to one trillionth of the universe,' or whatever the ratio is."

"So, what would this electronic robot version of yourself look like?" asked Duncan as the sun set outside the twenty-eighth floor, igniting the sky in reds, oranges, and yellows. "How would it work?"

"Look, you could build a very efficient digital-biological converter that would take your brain or your structure and print you a new body," said Enriquez. "It's pretty clear you can 3-D print body parts," he added, a process in which scientists in that time used a jet printer–like device to lay down rows of living cells to build simple organic structures, with whole organs a goal for the future. "It's pretty clear you're going to be able to use whole genomes"—all of the ACGTs in your DNA—"to reproduce organisms and clone them," he said. "If you really wanted it, hell yes, you could upload your brain, who you are, and take it digitally to land on Proxima Centauri b, and then print your body [once you arrive] if you really want one, or you build a robot to contain the digital you. The core question here, of course, isn't whether you can print your body. The core question is: Can you download what's in the brain?

"What I'm talking about here, the actual machine or version of me, would be tiny," he added. "It wouldn't make sense to travel through space in something big. Yeah, maybe a spaceship with your digitized mind would be the size of molecules—macro-size. It wouldn't look like a body. It makes no sense to design big."

"Like an advanced version of the so-called Cloud we have now," offered Martinez.

Enriquez nodded as he tipped a bit more Glenfiddich into each glass. Outside, the last wisps of light were giving way to the night and the glow of the city.

"I see it as an individual super-sentient being that communicates with similar beings," said Enriquez. "I don't see it as a Picasso painting, where the foot is over here and the face is over here. You have these coherent, self-contained parts that act and think as individuals, that are part of the far greater collective." He underscored that people would need to retain their separate identities. "This is why these robots that we download ourselves into can't be, and shouldn't be, cookie cutter, manufactured to interpret every feeling in the same way and every color in the same way. Part of the essence of our humanity is that some of us will hear a song and love it and some of us will hate it. Some of us will see a landscape and focus on very different parts of that landscape. So, you keep the randomness, the quirkiness, the unknowingness, the curiosity. In my mind, if you program a robot that isn't curious, that's a really shit robot."

The three men on the twenty-eighth floor each took a swig and pondered that one for a moment, until Martinez spoke up.

"Could we have a community of digitized minds where the three of us could be having this conversation, except we exist as just energy, or code, or something like that?"

"Absolutely," said Enriquez.

"Would we still be able to drink scotch?" asked David Ewing Duncan, feeling a really nice buzz and thinking, *I kind of like being an unenhanced human just like this right now.*

"Why not?" said Enriquez. "We could download the essence of the taste and experience of drinking scotch."

"What about touching another person?" asked Duncan. "A lover, let's say? Could you still have the sensuality of skin-to-skin? Let's say you can do that virtually, like the scotch, but would that be the same?"

"You have to go back into what's happening in your brain," said Enriquez. "When you touch something, like your lover, or you see a great movie, that's not getting stored in your brain as the actual physical molecular contact. It's processed as electrical impulses and chemical impulses."

Martinez and Duncan asked Enriquez if he thought this digital human world would ever actually happen.

"I think we're going to see this happen, unless we do something really stupid: nuke ourselves, or climate change. Or maybe these technologies are how we survive something terrible by removing the need for us to live in a world that is radioactive or drowning in carbon. What we need to ask is: How in the hell do you make this system effective enough to prevent the extinction of not just this species but all subsequent derivative species?" (Presumably, this referred to all the digital and otherwise enhanced species that humans might create or self-evolve into in the future.)

And so it went as that evening in 2018 deepened around the little group of friends perched high above the city discussing humans going *Homo digitalis* while they polished off a bottle of Glenfiddich. Not long after this, Enriquez launched a blitzkrieg of talks, books, radio shows, and venture investments. Eventually, this led to engineers and entrepreneurs actually building the machines that Enriquez dreamed of, technology that in the future became the basis for choice #1: *Homo digitalis*. Yet as those of us living in the future are well aware, Juan Enriquez didn't merely talk the talk. To prove *digitalis* was possible and desirable, he arranged with some of his tech buddies to be one of the first people to actually have his mind and essence downloaded into a computer. This happened after the usual experiments and tests to establish safety and efficacy, of course—even if the WFDAA (World Food, Drug, and Augmentation Administration) had not yet approved *digitalis*-tech for humans.

The last time we heard from Juan Enriquez—those of us living in the future who selected choice #2: *syntheticis*—he had uploaded his digital essence into the circuitry of a macro-starship and was heading to Proxima Centauri b. His tiny craft was fully equipped with programming and hardware to allow him and his fellow macrodigi-nauts to be in a state of constant learning and questioning and curiosity for the journey's hundreds of years. (Spaceship propulsion systems remained really slow even in the future.) As Juan Enriquez's macro-ship departed our solar system, he texted to his *syntheticis* friends, including Rodrigo Martinez and David Ewing Duncan, a final message:

ADIOS, AMIGOS! TALK TO YOU IN A FEW HUNDRED YEARS, GIVE OR TAKE. I'LL PROVIDE THE SCOTCH!

Wow, this sounded so thrilling to those of us in the future who chose *syntheticis* that we wondered if we had made the right choice. By then, however, *Homo syntheticis* types were moving pretty fast themselves toward enhancing smarts, strength, durability, and all the rest, and to allowing safer travel in space. These processes were heavily inspired by Harvard geneticist George Church, the champion of choice #2: *syntheticis*.

Back in the early twenty-first century, George Church, at six foot five inches, towered over most unenhanced people. He also had a long, gray beard like a wizard from the Harry Potter books or *Lord of the Rings*. Some people back then actually wondered if Church was a magician, given his steady output of mind-bending science that poked and prodded DNA and delved into the biological, chemical, and, yes, digital secrets of life.

In contrast to Juan Enriquez's ideas for *Homo digitalis*, Church in the early twenty-first century was devoted to keeping a flesh-and-blood body. "I'd like my robot to look like a human being," he said. "I like humans; I like myself as a human, although I think we can make healthier, more diverse, and more skillful humans." This made sense for Church, a biologist, since he best understood and trusted the idea of bio-enhancing humans and was suspicious of going fully mechanical or digital, in part because many of the robots and AI systems of his era were still kind of clunky. "We've got robots and computers that are supposedly beating humans at cognitive tasks," said Church. "I'm not impressed. They're winning at Go and *Jeopardy!*, but they're using one hundred kilowatts of power to do that, and they're competing against humans that are using maybe one hundred watts for the whole body and twenty watts for the brain; it's not even a close contest. Also, the humans are doing a whole lot more than focusing on the game. They're computing all kinds of things. They're computing their relationships; they're computing about justice, and ethics, and so forth.

"With robots," he continued, "it seems like you have to really spell things out for them. I think we would like to be able to communicate with them the way we do with children. You tell a child, 'Be a good boy.' You don't say, 'If you are going down the street and you're going to step on a kitty cat, don't step on the kitty cat.' You might give them a few examples, but then the child generalizes, right? Maybe machine learning will change this, but certainly for ethics, it seems like there will be a period of time where there will be uncertainty as to whether robots are thinking about ethics the way we think about ethics."

From his lab on the second floor of the massive glass-and-steel New Research Building at Harvard Medical School, Church back in the ERE (Early Robot Era) labored to make the *syntheticis* vision real in one of the largest and best-funded academic biomedical laboratories in the world. Where, among other things, in 2017 he started to synthesize the first human genome built from scratch, designed on a computer and assembled using small stretches of custom-ordered DNA. What he was working on was the difference between *reading* DNA, which is what the Human Genome Project accomplished when a rough draft was completed in 2000, and *editing* DNA, which back then was all the rage, and actually *writing* sequences of As, Cs, Gs, and Ts—a distinction that is very much like an author writing a complete book rather than reading or editing one already written.

"We want to synthesize modified versions of all the genes in the human genome in the next few years," Church said back then, standing behind a narrow lectern in his office that he used like a desk.

Note that he said *modified*. This meant that Church wasn't just synthesizing a copy of a human genome. He was redesigning it to be enhanced. His team did this by replacing select nucleotides—those ACGTs of life—and changing, say, a T to an A; or a C to a G in a process called "recoding." At the time, Church was "recoding" a human genome to make cells using his synthetic genome resistant to viruses, "like HIV and hepatitis B," he said, and possibly the common cold. Other possible recodings, according to Church, included making cells and organisms highly resistant to radiation in space; or, in the future, making humans recoded to have super strength or an IQ of 10,000—assuming we can figure out the genetics behind these traits.

"Imagine a good year for Einstein—1905," said Church, citing the year that Einstein came up with the theory of relativity. "Imagine that this was an average day for an average person," thanks to recoding. "That would be pretty incredible; it's hard to even imagine. There's no law of physics that we're breaking, because Einstein proved it is possible for one person having a really great year. Maybe everybody could, and maybe that would be the mean rather than the exception. If that's the average, then imagine what the edge of the bell curve is for that population?"

Church was also using synthetic biology to grow mini-organs. "We're making organs in dishes now," he said, "and there's no reason why we couldn't make brains in dishes. Those brains are not going to have the limitations we have. For example, their skull size won't be fixed. So far, they've been smaller than the organs inside of us, and they're not anatomically correct, which is why we use the term '–oid,'" as in "organoid brain" for a small, incomplete brain that scientists grow using stem cells, special cells that scientists can engineer to grow into any cell in the body. "We have grown about a dozen different tissues and blood vessel components; nerves, and glial cells that help service the nerves; cardiac and skeletal muscle; and almost everything you can imagine."

Another Church project back then was using DNA as storage. Church wanted to replace zeros and ones and silicon wafers with nucleotides (ACGT) as the code, while using the double-helix super-structure of sugars and phosphates as the medium to store the data. In 2012, he converted the words and images in a book he cowrote into genetic code. The book was called *Regenesis: How Synthetic Biology Will Reinvent Nature and Ourselves*, coauthored by author and philosopher Ed Regis. Church's team even converted a movie clip from digitized film to DNA and stored it in the DNA of bacteria, and then retrieved and replayed it. "Storing data in DNA has three advantages," explained Church. "One is that it's tiny, about a million times smaller than any other storage media. And it has a good longevity track record, lasting at least seven hundred thousand years." For instance, in 2013, intact DNA had been discovered in the Yukon, preserved in the femur of a horse that lived 560,000– 780,000 years ago. "Third," he said, "DNA has zero energy consumption during storage," which is amazing given all the power it took to store all those zeros and ones.

One day, according to Church, DNA might be used to store your experiences, too. "Can you imagine a cell that you put inside your brain that could be holding your memories?" he said (see "Memory Bot"). "What about an organoid brain that you just kind of grafted into your brain that you would program a certain way? There's plenty of room in there. And the cells take up a lot less space than electrodes." He posits that all human knowledge back in the twenty-teens could have been stored in DNA filling an average-size room.

If this wasn't enough, George Church was also working on technology that he claimed would reverse aging. "We're working on

6

reversing aging by harvesting a pretty vast and rigorous literature on things that cause longevity or other desirable aging consequences in animals, genetic and other factors," he explained. "We want to see if you can turn this into gene therapies that you would apply to an adult [nonhuman test] animal, and then [if was safe] move it into humans." Also, his lab was learning why some individuals, both animal and human, live healthy lives much longer than others and whether they can identify mechanisms that could be used to slow aging or maybe even reverse it. "The idea is to recode a person's DNA to make them younger," he said.

True to form, George Church started a company dedicated to reversing aging called Rejuvenate Bio—a business we are all very familiar with in the future if we chose to go *Homo syntheticis*. Back in the late twenty-teens, however, Rejuvenate Bio was still very small and mostly in stealth mode as the scientists there worked to reverse aging in beagles—really!—although there was no guarantee back then that their methods would work in either beagles or people.

Thus, in the future it came to pass on a certain Thursday in April at exactly noon GMT, each human then alive made the Choice. Various philosophers, ethicists, artists, tycoons, politicians, scientists, movie stars, sports stars, psychics, business leaders, and priests had weighed in with their thoughts. They offered advice and, in some cases, also created and sold various products that complied with the global guidelines to get people prepped for how they would become *syntheticis* or *digitalis*, and what to expect.

Polls taken before the Choice suggested that people were almost equally divided about which future they would choose. Which in fact is how the Choice actually came out, with 50.1 percent opting to go *syntheticis*, and 49.9 percent opting to go *digitalis*. This was possibly the first time in history that a poll was correct, even one augmented by a supercomputer AI brain. What the polls couldn't predict, however, was which option specific individuals would choose. In some cases, people had made their preference known, to friends and loved ones, while others waited until the last moment. This led to wrenching scenes on news vids of mothers choosing *syntheticis* and their teenage children going *digitalis* (parents made the choice for kids under thirteen years old). Or husbands choosing *digitalis*, and wives *syntheticis*.

Not long after this was when Juan Enriquez and other *Homo digitales* took off in their macro-starships for Proxima Centauri b and to other planets even farther away. At the same time the *synthetici* started their upgrade processes to implant and connect brain organoids and to radically recode their DNA and to reverse aging in humans just like George Church had done years earlier in beagles. Mostly, the new mini-brains were developed and manufactured by companies cofounded by Church, the main products being the Brainlet® series of neural upgrades. Brainlet® 1 was the first version, followed by Brainlet® 2, Brainlet® 3 and 3Plus, and so forth. Some of these organoids required an enlargement of the skull, which led to the fashion industry creating whole new lines of stretch hoodies and oversized hipster stocking caps.

Other enhancements included designer genes and cells that allowed the *synthetici* to travel safely into space. As the years and centuries passed by, companies created products that allowed humans to breathe without oxygen, even in the vacuum of space, or to explore the bottom of the deepest parts of the ocean without their bodies being crushed by water pressure one thousand times greater than at sea level. One enterprising *syntheticis* start-up in the distant future created upgrades that allowed customers to swim in the methane lakes of Titan, although it took several years in bio-stasis for even an enhanced bio-human to travel the vast distance to get there (see "Mars [Daemon] Bot").

Before we knew it, several hundred years had gone by since humanity made the Choice, when billions of people either went synthetic and remained mostly on Earth and the nearby planets, or they digitized their minds and their essence and departed for the stars—and then surprised us when they stopped communicating with us. This vanishing of the *digitales*, of friends and loved ones, made the *syntheticis* folks upset and sad for quite a while. Yet as time went by, people moved on.

Then, at exactly noon GMT, on another Thursday in April several centuries years after the Choice, we received the first messages from Proxima Centauri b telling us that the first macro-starships had arrived at last. They hadn't communicated before because they were so enmeshed in their own digital worlds learning stuff, but now they were ready to reconnect. Everyone was thrilled to hear from Juan Enriquez himself and others among the first *digitales* to leave Earth all those years ago.

Enriquez sent a series of detailed messages describing how the voyage had been long and difficult. But these intrepid former humans had learned a great deal with centuries to ponder things, including how to create (macro) robot versions of themselves that looked eerily like they had when they were flesh and blood. Enriquez sent back pics of himself that showed his robotic white hair and beard that were still kind of wild and professorial. They also sent images of their alien but hauntingly beautiful new home, with its red-tinted sky as the planet orbited Proxima Centauri, a red dwarf sun.

Not long after this, Enriquez sent a personal vid to his two friends, Rodrigo Martinez and David Ewing Duncan, both of whom were now fully equipped with brain organoids and other synthetic enhancements. The short vid showed Enriquez's new robot self toasting them with a glass of scotch, the taste of which he had beamed to Martinez and Duncan in digital form for them to download and synthesize. And Martinez and Duncan had to admit, the single malt beamed back from Proxima Centauri b from digital Juan was delicious—and gave them a nice synthetic buzz.

TOURIST (EVOLUTION) BOT

Right now, the future version of you might be thinking, *Boy, do I need a vacation!* But where to go? You've already visited thousands of places virtually and quite a few in reality. You're wondering where you could travel that's fresh and new. That's where Evolution Bot comes in, a sophisticated Bio-Quantum Molecular Dislocation System that takes you on tours through critical events in Earth's evolutionary history over the last three and a half billion years. Those moments in the past when life on Earth experienced dramatic environmental shifts either locally or globally—the coming of an ice age; the change in food sources in an isolated valley; the arrival of a new predator on an ancient savanna—that caused some organisms to die off, while others survived because they and their DNA adapted to save them and to pass down mutations and other molecular changes to their progeny, including, eventually, humans.

"Imagine that you could go back and experience when the first creatures crawled onto land and developed the capacity to breathe in air," said the man who first dreamed up the idea of an evolution bot way back in the twenty-first century, the designer and entrepreneur Rodrigo Martinez. He also came up with the idea of using Evolution Bot—"Evo Bot"—to allow people in the future to visit pivotal moments in our evolutionary past as tourist destinations. "Or you could be there when the meteorite hit the Earth that ended the age of the dinosaurs," which probably caused a decades-long nuclear winter sixty-six million years ago that may have led to the rise of mammals that survived to eventually dominate the Earth. "What if you could experience what this was like from the point of view of a lemur," said Martinez, "and find out: How did lemurs and other mammals adapt to survive the meteorite and proliferate? How did that feel?"

Hey, you have to admit that taking trips backward in time to visit pivotal moments is certainly, ah . . . different. Different from, say, lying on a virtual beach on Titan reading the latest Baldacci-Nova thriller or taking the family once again on a moon excursion to Disneyland Luna.

The actual process of evolution works, of course, primarily through "natural selection" and "survival of the fittest," theories articulated by Charles Darwin that species over time either survive or don't based on their ability to adapt to changes in their environment. Mechanisms of evolution include individuals in a species that have genetic mutations that favor their survival that are passed on to their offspring and their offspring's offspring. These can take several generations to effect change—like the capacity of humans to stand erect, which probably developed over thousands of years from apes that mostly crouched, which meant they couldn't easily see over the top of grass growing on the savannas to spot a saber-toothed tiger coming their way. Many of the major evolutionary changes over the eons have been recorded in sequences of DNA in every organism, including humans. Some of these sequences remain vital to the organism, others are mostly dormant, but all represent genetic echoes from the past, like sequences of DNA that helped us breathe underwater when our ancestors lived in the sea. These stretches of DNA in humans were deactivated millions of years ago and many have disappeared, but trace sequences remain to tell the story of how our ancestors evolved to crawl out of the oceans and adapt to breathing air.

Another major evolutionary mechanism is called epigenetics, which allows organisms to adapt much faster than genetic mutations. Epigenetic changes occur when an organism reacts to a meteorite or volcanic eruption—or an infection by a virus—by turning on and off genes that help it survive. They then pass those changes on to their descendants. One example of epigenetics came in 2018, when researchers discovered that, in World War II, mothers who were pregnant when the Nazis cut off food shipments to Holland during the winter of 1944–45, bore children who were smaller than average. Later in life, these former famine babies also suffered from high rates of obesity, diabetes, and schizophrenia—maladies that, in some cases, were epigenetically passed down to their children and grandchildren. Scientists also have theorized that epigenetics can produce beneficial effects in children when genes are turned on or off as a response to positive reinforcement and nurturing from their parents, meaning that some epigenetic changes might be helpful.

Really, it's a little crazy that Rodrigo Martinez way back in the twenty-teens thought up the idea of turning the ancient record of evolution that's stored in our DNA—the sequences of code that either mutated or were epigenetically turned on or off—into a travel business. Especially since the technology didn't exist back then and took decades to develop in the future into what became a kind of virtual evolutionary time machine.

"The engine for this travel back in time will be the record of mutations and epigenetic changes that are preserved in our DNA," said Martinez back in 2018, "and this will happen in a robot you will wear like a body glove, connected to your neural system, you and your friends and family that will travel with you. Together you will go to these moments in the past where these evolutionary changes happened. Maybe you'll go all the way back when groups of early cells were floating in water, and they experienced some sort of environmental change. We don't know how to do this yet, but in the future when we understand the role of specific mutations and epigenetic changes better, it's possible that we will be able to create a kind of experiential superhighway all the way back to the beginning of life, which could allow you to travel to any point in the history of life."

Martinez got his idea for Evo Bot from his love of the sport of free diving, where divers hold their breath and swim down from the surface of a pool of water to one hundred feet or more without a tank of air. In the twenty-teens, Martinez was a free-diving aficionado who loved to plunge into freshwater holes called cenotes that dot the Yucatán peninsula in Mexico. "What I feel when I'm free diving, and I'm floating in water at one hundred feet, is something that I can't express in words," said Martinez. "It's primordial, more than what we call an emotion or an expression. Maybe what I'm actually feeling is an evolutionary echo of what other cells have done in the past. I don't know. It might be metaphysical. It might be spiritual. I'm just feeling something that I know connects me with something really, really, really far away. I don't know what to call it. Maybe it needs a new name."

Martinez was sure that his love of free diving as a spiritual ex-perience might have something to do with an evo-echo from the deep past, from organisms and maybe human ancestors that really loved being in the water. "It's an expression of my DNA," said Mar-tinez. "I close my eyes and I try to make a connection with the en-ergy around me. Maybe it's God, or something like God. You can't go and search that. You have no way to do that. When I meditate, I have this sensation of connectiveness with the universe; maybe that is also an evolutionary connection. But we just have no way to be able to tap into it."

So how would an Evo Bot work?

"I think it's neurobiological," said Martinez back in 2018. "Or it feels biological. Right? Imagine that I'm connected to this thing. It's like a sleeve, or a pod you wear, maybe you're sitting in a special chair and you're wired in, or you're connected biologically with wet-ware. And we can program into this robot, or whatever it is, to start to give me the sensation of the first set of molecules floating in wa-ter. I feel suddenly this moment of existence that was not there be-fore. Right?

"It's partly virtual," continued Martinez, "and you see it and feel it. You're a part of it, but you're also watching it like some sort of immersive movie. What I want is to feel the expression of the gene or the genes myself. I want to go and sense that expression. Like what happened to the dinosaurs when the asteroid hit, what hap-pened to the lemurs." There is also this possibility that evolution and epigenetics could be harnessed to make positive changes in hu-man genes. "What if you are put in a positive situation, and you can pass that emotion or feeling down? What if we can prove in fifty years that there is an epigenetic or evolutionary echo that we could set up for our descendants that would give us new models to think about ourselves, and life, and possibilities?" Or what if projecting positive epigenetic or evolutionary "memories" into the future could become a vehicle for eliminating diabetes or schizophrenia?

It took longer than Martinez hoped it would, but eventually his Evo Bot became a sensation in the tourist industry of the future, with special deals, discounts, and "evo points" awarded by Expedia Bot, Kayak Bot, and other neuro-linked travel sites. People loved exploring evolutionary moments culled from their DNA, those pivotal points in evolution where ancient trilobites developed to swim in primordial oceans, birds first learned to fly, and on and on. Powered by the latest Bio-QMDS, Evo Bot offered the opportunity for a human tourist in the future to slip into a bio-pod sleeve that was controlled through a dashboard that allowed you to input the experience that you wanted to have, which became a blend of elements both virtual and real.

There was, however, a wrinkle that developed. As one might expect, many of the pivotal evolutionary moments in our past were quite disturbing, what with meteorites killing off entire categories of life, and starvation during the Second World War, and all the rest. Turned out that some tourists in the future couldn't take the trauma. Just like some people who go to Disneyland Luna can't deal with the g-forces of a monster roller coaster plunging straight down for seven miles, even if the moon's gravity is only one-sixth that of Earth's. That's when Martinez and the Evo Bot team developed a ratings system, kind of like the old movie ratings system, with a rating of G describing epigenetic experiences that were benign and enlightening and kind of fun, like when the first animals emerged from the sea to breathe in air on land. There was PG for evo episodes that were a bit more challenging for small children and for the squeamish. And so on, with cataclysmic, near-extinction events designated as D+, for "Disaster +."

Despite this ratings system, Evo Bots were nearly shut down when a group of scientists gathered evidence that in some extreme cases they were actually creating fresh epigenetic and evolutionary changes in the DNA of some tourists, changes that might impact them adversely later in life, or their offspring and offspring's offspring. This led Rodrigo Martinez's company to offer to pay for medical procedures that reversed these epigenetic and evolutionary stresses from being passed down to an evo tourist's progeny.

For a little while, Martinez and his Evo Bot engineers had to dial back on the D+ cataclysms of the past and offer more benign evolutionary pivots. Like when the first Cro-Magnons had sex with Neanderthals, which was surprisingly lovely, sensual, and pleasant (age 21+ only). Or when one of our ancestors experienced the first-ever glass of wine—a proto Pinot Noir, as it turned out. Can you imagine all the amazing genes that began to express and turn on and eventually mutate with each sip over the generations as the first fermented grapes were poured out into a shell or sheep horn or whatever these ancient people used before stemware was invented?

That's when Martinez introduced Evo-Epi Bot II, designed to capture happy epigenetic and evolutionary moments in the past, with an upcoming upgrade that Martinez's company said would allow you to project positive epigenetic moments into the future for, say, your great-great-grandchildren. These evo-epi experiences could be evo-epi engineered to make your descendants happy and well balanced, or strong and smart, and so forth.

Then came that strange day when the millions of people enjoying various Evo Bot experiences suddenly got a really weird feeling en masse. It was a powerful sensation that someone or something was watching them. But who? And why did it feel so . . . creepy, but also kind of exciting?

Many believed that we were sensing evo tourists from the deep future who were traveling along their own evolutionary highway back in time to visit us as part of their own evo vacation (see "God Bot"). Remarkably, Rodrigo Martinez back in the early twenty-first century actually had predicted this might happen, too. "The other thing we might find out when we get to a meta-evolutionary level is that we are actually an evolutionary expression of some other being," said Martinez. "We are the evolutionary experiment. Just like we're thinking: 'What would that first cell feel like?' Right now, we are the early evolutionary version of something else thinking: 'What would it be like to go way back to experience being human?'"

In the future we never did figure out the identity of these shadowy observers. At first, their appearance in the middle of our evocation freaked us out. Then we tried to communicate with them, which failed. Finally, we concluded these strange entities were probably an advanced version of post-humans living millions or billions of years into the future who were indeed exploring their own epigenetic/evolutionary history, which led back to us humans.

Some people, once the shock of being watched wore off, concluded this was actually kind of cool. Others worried that these future beings might try to kill us or turn us into pets. Others worried about privacy, since unknown entities from the distant future were kind of like trolls hacking into our lives. This led to demands that Rodrigo Martinez shut down his evo-cation business and that evo tech be banned. The World Congress even held tele-vid hearings where Martinez was grilled for three straight days, with politicians asking him if evo tech was really as safe as his company claimed. In a narrow committee vote, Congress decided not to ban the tech while we waited to see what would happen.

Months went by after the vote with nothing bad occurring, which eventually meant the whole banning thing went away as most of us got used to the bizarre sensation of entities from the distant future hovering just beyond our consciousness while we vacationed in, say, the Cretaceous Age sixty-six million years ago, or when bacteria ruled the Earth. Lots of people ended up liking these apparitions being there to share their vacation, even if it remained kind of frustrating that we couldn't communicate with them.

Sure, people kept trying, until we realized that this was silly. Because if these entities really were superbeings from millions or billions of years in the future, then us trying to talk to them would be like a single-cell organism from the early goo-ponds of life three and a half billion years ago trying to talk to us.

GOD BOT

hen the moment arrived in the future when linear time ceased to exist and the fourth dimension became moot. When we shifted from an existence where events occurred in order, one after another, to one where everything occurred at the same time. This happened when God Bot arrived at last—a bio-quantum infinity cascade *umwelt*-system (see "Wearable Bot") equipped with the most advanced multidimensional-neural-uncertainty field ever built. One that could experience the beginning and the end of the universe, and everything that happened in between, in a continuum with no actual beginning, middle, or end. (Don't even try to get this if you haven't learned at least the basics of bio-quantum infinity field theory, which is a freshman introductory course in the universities of the future.)

When we named this ultimate robot "God Bot," it had nothing to do with religion or worship. It wasn't the bot equivalent of a benevolent god with a long gray beard sitting on a golden throne, or a goddess with long, flowing hair and a crown of daisies living in a celestial treehouse. God Bot did not live in a burning bush or touch us with epiphanies of love and grace. God Bot was not about sin and redemption or about heaven, abundant love, empathy, or reincarnation. And it certainly wasn't an entity that the few rabid true believers who were still around in the future could use to justify killing and torturing infidels who didn't worship the right God Bot or the correct sacred algorithms.

One of the most important rules, or realities, that we learned from God Bot was that the universe really just *is*. It works according to a set of specific physical rules that govern the formation and destruction of celestial bodies, and dictate gravitational and quantum forces, and so forth. But it is not imbued with an underlying and irreducible code of moral rights or of empathy. The universe doesn't shed a single tear when black holes suck in and crush all nearby matter, stars die in supernova flameouts, and icy comets traverse the vast, empty reaches of space all alone. The universe couldn't care less about the rise and fall of advanced civilizations on various worlds or when intelligent life snuffs itself out because of some stupid doomsday weapon they built. Or when a civilization is daft enough to allow carbon to build up in their planet's atmosphere to the point that no one can breathe and the world that sustained and nurtured them dies.

This lack of compassion from the universe was a huge disappointment to people longing for meaning and purpose in the vast ether of space and time. Good and evil kind of stuff, and perhaps a universal system that rewards people (and robots) who are good and ultimately punishes those who misbehave. Other folks actually embraced the continuum and infinity of a heartless and soulless universe as validating what they believed was the true meaning of existence: relishing the wonders of the universe as revealed by God

Bot. These devotees hung on every quantum dot, Higgs boson, and new dimension that God Bot revealed. They welcomed a universe that to them was a vast stage in space and time, where an infinite cast of celestial players engaged in endless dramas, large and small, that played out according to the rules of physics. The birth and death of galaxies, stars exploding, and whatnot. In its own way, this was as exciting as the traditional myths, fables, scripture, and parables that people once told—and were thrilled by—that starred humanlike gods. But what really got these God Bot devotees going was the idea that the physical and dimensional rules of the universe were kind of like the stern, unyielding rules promulgated by old-style religions where God was a hard-ass who demanded you follow His or Her rules—parameters these people seemed to need to avoid feeling lost in a godless universe.

Once, long ago, back in the ERE (Early Robot Era), the launch of a god bot might have been hailed as the arrival of "the singularity," that magic moment when humans and machines were supposed to become one and propel us into a dramatically different relationship with the universe—a situation that some thought would go really well, and others thought would cause the extinction of *Homo sapiens*. But no one knew for sure, in part because there were so many definitions of the singularity, the simplest being a moment in time when something singular and unprecedented happens that changes everything. Under that definition, the appearance of God Bot certainly qualified, a machine that could know everything that happens in the billions and billions of years that exist between the birth and death of our universe.

God Bot, however, was, for most people, a teacher and a guide to the secrets of the universe and to the continuum of time and space. Some folks were so eager to learn and to immerse themselves in the continuum that they begged God Bot to allow them to fuse with its supreme mind, to become one so that they, too, could understand how everything worked. Physicists were the most eager to learn and to fuse, having spent millennia trying to devise clever theories using equations and proofs about how the universe worked. A few philosophers also wanted to better understand God Bot's revelation that the universe lacked the sort of moral underpinning that most of them had staked their careers on, even as they were skeptical that this could really be true. These so-called fusioneers were almost delusional with joy when God Bot provided them with the neural upgrades required to comprehend the universe. This was before they fully merged with God Bot and the universe and one another, although even after they fully fused, they weirdly were still able to hang out with the rest of us in linear time at places like the Aspen Ideas Festival, TED, and San Francisco Giants games (God Bot didn't much like the Dodgers).

Most of us who didn't partake in the fusion with God Bot—and preferred to tenuously hang on to at least the illusion of linear time and the existence of traditional mores of good versus evil—could still appreciate God Bot's vast celestial knowledge. This was made easier by a raft of companies that cut deals with God Bot to offer a variety of temporary fusion "experiences," from basic and deluxe to "the ultimate," a God Bot app that allowed people to visit all corners of the universe, past and present, like tourists. This included being present at the Big Bang, when the first particle of energy appeared in the void, and also the very moment when the universe ended and crunched back onto itself. (Physicists from the ERE might be interested to know that the theory proposing that there were infinite big bangs and big crunches happening one after another, plus the multiverse theory of an infinite number of universes, ended up being true—something God Bot will be happy to explain once it's invented.)

For some people, however, even the most basic God Bot experience was too intense. These humans preferred to stay home to watch others take interdimensional excursions across time and space. A few complained that all this cosmic hustle and bustle made them dizzy, even nauseated. For these folks, God Bot provided mini-dramas and series about the universe's spectacular stories, all neatly packaged with beginnings, middles, and ends, in the manner that narratives used to be told before the continuum.

God Bot as an idea first appeared in the early twenty-first century as a yearning by a theoretical physicist named Brian Greene. He was a professor of physics and mathematics at Columbia University, where he studied things like mirror symmetry, conifolds, orbifolds, and Calabi-Yau manifolds—head-scratching concepts and methods for most of us that physicists used to explain shapes and dimensions in space. "He also described the flop transition, a mild form of topology change, showing that topology in string theory can change at the conifold point," said Wikipedia back then, in its profile of Professor Greene, terms that really seem kind of made-up, but they're not.

Thankfully, Brian Greene helped explain things to non-physicists by describing these terms and much more on popular television series and in bestselling books with titles that made physics sound beautiful, like *The Elegant Universe* and *Icarus at the Edge of Time*. Mostly, though, he spent his time pondering the cosmos and scrawling dense mathematical equations in an attempt to understand the continuum long before God Bot technology existed. For instance, Greene was among a small group of physicists who argued that there might be eleven dimensions in the universe, seven more than the usual four dimensions of length, depth, height, and time. Some of these extra dimensions, said these physicists, ran through the universe like pieces of string, which was why Greene and others dubbed this "string theory."

Still, Brian Greene and his colleagues were frustrated to have to describe ideas like string theory using dense mathematical equations that only a few people understood and that scientists had a hard time validating in the real world. It was this vexation that prompted Greene one day in 2018 to imagine a robot that could help him better understand and prove his theories while explaining deeper truths about the universe both to him and to non-physicists—a concept that became God Bot.

Greene didn't imagine a fully realized God Bot all at once. He first had to think about it and to talk about what he was looking for. "I would love to interact with some kind of AI," he said in a phone call from his office in New York City in 2018, "some kind of future bot that had the capacity to *really* take in our deep understanding of the world emerging from quantum physics; and emerging from relativity; and emerging even from the strange ideas of string theory and extra dimensions." A man with quizzical but languid eyes and dark, short-cropped hair, Greene spoke really fast, lining up arguments and thoughts as if his mouth could barely keep up with the speed of his ideas.

"I'd want a bot that could incorporate this knowledge in a way that was visceral and intuitive and would allow me into these worlds," he said. "That also would allow me to not just see these ideas and equations, which I struggle to wrap my mind around right now, but to understand how they would actually work. What if this bot could bring me into these strange new realms, so I could understand them as deeply and intuitively as I understand everyday life? That to me would be the ultimate move in the direction of really what I've spent my professional life trying to do, which is to grasp reality from radically different perspectives revealed to us by research into the fundamental laws of the universe."

That sounds amazing, assuming you understood what Brian Greene just said. Even if you didn't, let's keep going with the hope that things will become clear or that God Bot will come along to explain what Greene was talking about.

"To me the most vital part of what makes life interesting," said Greene, "is to see reality through different lenses. That's what we physicists have done as a group for hundreds of years. With the Newtonian view of the world, which is very mechanistic and very straightforward, it aligned with our everyday intuition. Then came the Einsteinian revolution, with relativity changing our views of space and time, which required that we learn those equations and gain an intuition for them. Next came the quantum revolution with its new equations, which completely changed our sense of how the world evolves over time, a completely new intuition and perspective of the world.

"I feel that our intuition is largely informed by the unconscious processing of the human mind," he said, "a process that we don't entirely understand but informs us and becomes something we feel and trust. Think of the great musicians, the great pianists of the world; they have embraced their instrument at such a deep level that they don't need to think about the movement of their fingers. That's an unconscious process that ties into some of the deepest creative output of the species. I have spent my life gaining an intuitive understanding and a subconscious and a deep unconscious understanding of mathematical patterns. You show me certain mathematical patterns and I immediately know what they are; I don't have to think about it. Much as a great pianist can look at a piece of music and immediately the fingers start to fly, they don't have to think about it, it's right there.

"It takes a lifetime to get to that place," he said. "If the bot, the AI, that we encounter can somehow augment our unconscious, so that we don't have to spend a lifetime gaining that capacity, then all of a sudden you could look out in the world and immediately feel mathematical patterns. You could feel musical patterns, you could feel artistic patterns, you could feel linguistic patterns. You could feel the kinds of patterns that it takes the expert a lifetime to acquire. If we can gain that unconscious quality—wow! That would be amazing.

"What if I could hold the bot's hand and just through that contact be able to acquire the unconscious processing that the bot presumably has acquired? It's able to learn much more quickly than I can. It presumably has far greater longevity than a human life-form, and therefore it could embody hundreds of lifetimes' worth of deep experience that has been absorbed at a level that now sits beneath consciousness, one that sees the world by being informed by its deep intuition."

Okay, so we have this super-intuitive robot. We don't yet know what it looks like, except that it seems to have at least a hand (and maybe fingers?) for Brian Greene to hold while its super AI quantum brain imparts its deep understanding of the universe into Greene's subconscious, so he could perform physics like André Previn played the piano.

But besides having a hand, what does this robot look like?

"It almost doesn't matter what it looks or feels like," said Greene. "What really matters to me is what's inside. First, I would need this robot AI to be conscious. It needs to have a sense of self, it needs to have a sense of experience in the world, and it needs to be able to look out at the world, guided fully by the mathematics that modern science has developed to take us to the next level of understanding. And not just see that as equations but embrace it as a visceral sense of how the world is put together. And bring me into that, too, and be able to communicate that.

"Now, I don't know if it communicates through words or touch or feel or just kind of hooks me up," Greene said, "so I can see the world through its conscious experiences. But that to me would be the radical move, in terms of a depth of understanding of how the world is put together."

This is where the bot Greene was imagining began to sound god-like. Again, not like the god in the Bible or the Koran or the Torah, but one that understands the deepest textures of time, space, infinity, and the universe and can magically impart this ultimate knowledge to Brian Greene almost instantaneously—and to other people who right now aren't able to understand even a tiny fraction of what theoretical physicists are saying with all their crazy equations.

"I can't help but feel that if I or other people who think about this stuff could have these experiences," said Greene, "then I think we would gain a deeper sense of whether our abstract mathematics is right or whether it's wrong. I think this would be hugely facilitating because there's a barrier right now. Most people cannot participate in this exploration because they can't wrap their minds around the ideas, largely because they don't have the technical training, they don't speak mathematics, they haven't learned all the necessary background material to intuitively follow the formulation of these cutting-edge ideas.

"Let me give you an example: if you were standing with me right now, and I was to pull out an apple and toss it to you, you would immediately move your right hand, if you're right-handed, and catch it. Which means that you figured out the Newtonian trajectory of the apple, and you knew exactly where to put your hand to intersect that trajectory. That's true whether or not you know Newtonian physics, whether you remember F equals ma [force = mass × acceleration], and all the basic equations necessary to calculate that trajectory. It's intuitive. It's practical.

"Later theories of physics aren't so intuitive or practical like this. That's because our early brethren who were walking along the savanna thousands of years ago and stopped to contemplate quantum mechanics and black holes, they're the ones that got eaten. So, there's been an evolutionary pressure selection against thinking about anything beyond the laws of classical, Newtonian mechanics, because you want to dodge that rhinoceros that's charging you. You want to know where to throw that spear. What I want is for our brains to be augmented to understand these other ideas at the same level. If, to survive, we needed to know quantum mechanics"—or relativity or string theory—"then all of us would have a quantum understanding of the world"—or a relativistic or string theory understanding of the world—"which this bot I'm talking about would take us to.

"How wonderful would it be if we could look out in the world with quantum eyes? Or relativistic eyes? Or multi-universe eyes, if that's correct? Let me give another analogy. Imagine that you live in a world where everything is 2-D and somehow you develop bots in this 2-D world that are able to see the world like you and I do, in 3-D. And they were able to have that information flow into the inhabitants of the 2-D world. That would be pretty dramatic."

But would such a device really deserve the appellation "God Bot"?

"It's deep and mysterious how people get to a certain place," said Greene, pondering this question. "And yeah, if you had the capacity to get there in basically any arena of interest, then yeah, maybe it does feel like you've been not just touched by God's finger, but maybe you've had the full palm put on your hand. You really are being able to experience reality in a way that perhaps only a godlike figure could."

But one had to ask: "Would God Bot really be for everyone?"

"My gut feeling is that this bot could be widely accessible," answered Greene. "It could even become the base tool from which humankind can create further things in the world," including businesses and products.

And wouldn't such a machine be so powerful that we might end up fearing it, kind of like we feared an old-fashioned god?

"I could imagine that by creating these entities we might ultimately see our role as humans diminished as we are subsumed by this entity," said Greene. "This is something that I wouldn't fear. Nor do I even think it's necessarily wrong if we become a stepping-stone in a certain kind of evolution, and there's a time when we're no longer part of the picture. That's fine with me, too."

Yikes! One might wonder if everyone would feel that way, although Greene suggested that this evolution would happen gradually as a kind of merger—which might make it less frightening.

"I do think it will be a merger," said Greene. "It's already happening, obviously." (See "Matrix Bot.") "And I do think that we'll be in ever-closer association with intelligent machines, at which point I think it's unlikely that we're going to be able to distinguish between what is human and what is AI. I think it will just be the new entity that we will evolve into. At some point that new entity may bear no obvious resemblance to where we began as biological, messy, physical beings. We will be in there, we will be entwined, just as ancient biological structures like the mitochondria are still in our cells today [a reference to the fact that the mitochondria structure that exists in most cells was at one time invading bacteria that stuck around and mutated to become a critical part of the cell's function]. It's been there for a long time, but we're not just mito-chondria," he continued. "I think that's fine if the human contribu-tion to this new entity is profoundly in the mix, but you can't really point to us any longer."

So, let's say humans become the mitochondria inside of some super God Bot entity. Then what?

"I think where it leads is to a hugely accelerated creative output and a hugely accelerated scientific output. I think this does lead to things like a deep understanding of the nature of space, the nature of time, the ingredients that make up space and time. A full and final accounting of how the universe came to be, a full and final accounting of what the universe is indefinitely far into the future. And a full and final accounting about whether ideas of a multiverse are coherent. Do they make sense? If they do, are they part of real-ity? I think those are the kinds of questions that we will want an-swered. At the moment, each and every one of those questions is murky. It's a total question mark on any of those."

But even if we could design such a system, would we actually be able to understand the universe at God Bot's level?

"I have little doubt that we would. Let me tell you one pattern that plays out, which is: each generation pushes our understanding further, adds to the corpus of material there that the next genera-tion needs to learn in order to push the boundary yet further. I've always been mystified, at some level, by the fact that somehow each next generation learns everything that I had to learn, plus the new stuff, and hopefully they're still young enough and energetic enough to push the boundary forward. With these sorts of AI assistance, if you will, the run-up time to be able to grasp the new understanding and internalize it in the way we are ready to go forward, that process would be short-circuited enormously."

Does Greene actually believe that he could experience, say, the eleven dimensions of string theory?

"Why not?" he said. "Again, that would be a thing that would be physically impossible, because if the extra dimensions were big enough for my body to fit within them, then my body would be unstable in that world. But if this is something where I am accompanying an eleven-dimensional stable being into this eleven-dimensional world—if that is what the bot is able to do for me, to allow me to see the reality through its eyes, and to feel the reality through its sensors and so forth—then yes, I think my intuition takes a spectacular leap."

Yeah, but was understanding quantum or string theory really something that lots of people *need* to know? Was there a practical reason to spend the time, energy, and money to build a god bot for this purpose?

"Yes," answered Greene, speaking even faster as he got more excited. "Why do we value Beethoven's Ninth Symphony? Why do we value *Starry Night*? It would make life all the more wondrous if we could experience reality face-to-face, so to speak, as opposed to the misleading Newtonian face that the world presents to us. I'd say that's perhaps one aesthetic reason. The other reason is that the history of science teaches us that the better grasp we have of reality, ultimately the better control we have of reality. And in that way, we can really change the world and have an amazing impact. Even though we're in a Newtonian world for the most part, the quantum nature of the world is why you have a cell phone. It's why you have a personal computer." (Quantum because engineers manipulate the electrical properties of silicon based on their knowledge of quantum waves of electrons.) "So, by virtue of grasping a little piece of the quantum world, we've completely changed everyday life. I can't imagine that the same wouldn't happen to an enormous degree with the kinds of explorations that we're talking about.

"For example, you take string theory—what is it? It's an attempt to create a unified theory that ultimately, we believe, will rewrite the understanding of space and time. This could provide us with a means to travel great distances in space. If we are able to manipulate time, maybe we would be able to more freely traverse the time axis of the universe forward, or possibly even backward, in time, who the heck knows? We're now sort of in speculation-squared mode. But the larger point is that deep understanding has always led to a deep control that is widely applicable, not just to the eggheads who study it. It can change life as we know it."

Not long after God Bot came online many years in the future— and changed everyone's perception of the universe, while radically transforming the economy and how people thought of themselves and the role of good and evil in a compassionless universe—Brian Greene disappeared. He was last seen on a live-feed interplanetary tele-vid in front of a vast virtual-live audience broadcast from a theater in New New York City, the capital of Proxima Centauri b. Greene had just arrived from Earth on a Timeliner, which was kind of like an old-fashioned jetliner, except that instead of flying just through space, you flew through space *and* time. People from all over the space-time continuum were watching as this future version of Brian Greene made a startling announcement—speaking, as usual, really fast.

Looking boldly into the tele-vid's camera, the future Greene (who like most of us had radically extended his lifespan by using long-life technology) explained that he and his physicist buddies had just discovered, while holding God Bot's hand, that the universe actually contains an infinite number of dimensions, not just eleven. This came after God Bot had taught them to intuit the intricacies of string and multiverse theories almost as easily as they intuited the Newtonian trajectory of an apple tossed their way.

Incredible! we thought, knowing that God Bot would soon tell the rest of us what the hell the future Brian Greene was talking about. That's when we watched Greene and a couple of his physicist friends deftly step through a multidimensional-neural-*certainty* field, as opposed to an *uncertainty* field. This was a brand-new technology that was a substantial upgrade to God Bot, one that would be coming out soon for everyone to purchase on Amazon Bot Neural-Prime.

Poof! Greene was gone, followed nanoseconds later by his friends.

Where did they go? Were they coming back? Had they merged with the new God Bot II neural-certainty field, becoming a cosmic version of mitochondria tucked into the vastness of the universe?

We didn't know, since no one else had the new God Bot upgrades used by the future Brian Greene. This left us in the dark about his notion of infinite certainty multidimensional travel—which honestly made us feel kind of grumpy after the sort of deity-like feeling that most of us had experienced since God Bot enlightened us. Suddenly, we felt like people who had been living in a 2-D world trying to comprehend 3-D. This left us feeling quite discombobulated. Some people even wept when the company that was developing the God Bot upgrades told us that their production team was experiencing some hiccups in manufacturing, which would delay the delivery of God Bot II. Amazon Bot insisted, however, that they would be available in just a short four to six months, guaranteed!

This left those of us in the future, after thinking that our tech had conquered time and space, even more frustrated. Kind of like long ago when Apple experienced delays in shipping the second-generation iPhone that everyone just had to have after being so wowed by version one. As always, people wondered what was taking so damn long, something that even God Bot was at a loss to explain, which made us wonder, *If we have really conquered space and time, then what is the friggin' holdup?*

IMMORTAL ME BOT

Something odd happened when we looked into our 3-D holo-mirrors, after billions of years of bio-enhancements and technologies that had eliminated illness, old age, and death. We realized that we had grown seriously tired of ourselves. How we looked, how we acted, what we thought about—tired, too, of experiencing all the things we could think of to do in the universe and across time (see "God Bot"). There was still stuff to do in the ever-expanding universe. But after optimizing and refining everything we could think of, millennium after millennium, and myrlennium after myrlennium (millions of years), and byrlennium after byrlennium (billions of years) with fads and styles coming and going, we eventually kept coming back to the same looks, sensibilities, and emotional settings, where everyone was beautiful, smart, athletic, charming, and funny.

Which was really boring.

For those who have watched the histo-vids from pre-ancient times, this was a little bit like what happened on Old Earth in a place called Beverly Hills in the late twentieth and early twenty-first centuries. People back then frequently indulged in something called plastic surgery, where human doctors—this sounds gross—actually cut into people's faces and bodies to supposedly help them improve their looks. They nipped and tucked to create luscious lips, tight asses, wrinkle-free faces, and bigger breasts for women (and sometimes smaller ones for men). These augmentations left everyone appearing more or less the same. Many opted to look like movie stars and celebrities now long forgotten. Sure, people back then might have dyed their hair green for a few days to rebel against the same beautifulness. Or they might have put a metal ring in their nose, or they got a tattoo of a snake, skull, or flower winding up their arms or on their butt. Yet in the end, most returned to the tried-and-true version of themselves that they most loved.

A few years after this, designer genes became available to edit into people on demand, which allowed a person to pick any face or body they wanted, although some people did dabble in being more creative. In the future we all have gone through phases of tweaking our faces and heads to look like a lioness or a giant panda. But these were passing fads and brief personal expressions, like green hair and nose rings were for pre-ancients. Wings were also a big deal for a while, until they became ho-hum after a million years or so of being able to fly like a bird. This may be hard for people back in the pre-ancient times to imagine. To them, flying like a bird would have been pretty cool. And it was, for five or six hundred thousand years.

The same thing happened with enhancements to our intelligence as scientists kept making us smarter and smarter. Eventually, everyone had an IQ in the millions, although weirdly all this intelligence didn't fundamentally change us once we had figured out solutions to most of the big challenges facing humans on Earth, and in space, and in multiple dimensions. All that brainpower actually made some people kind of go crazy with all the input and analytics swirling around in their giant, organoid-superenhanced brains (see "*Homo digitalis/Homo syntheticis*"). It made some people feel nauseated until they learned to throttle back a bit on their Brainiac XB-7500's super-smart settings. It also became less exciting to be one million times beyond Einstein brilliant when anyone could turn up their XB-7500 and be the most brilliant person alive, just like everyone else.

The monotony with perfection was a big deciding factor for many people a few billion years back, when half of all humans opted to fully digitize their brains and quit their flesh-and-blood bodies, in part to try something new (see "*Homo digitalis/Homo syntheticis*"). Soon after downloading their minds, these newly minted *Homo digitales* spoke to us in weirdly robotic voices that insisted it was so very cool to be digital and definitely not boring.

As if we hadn't heard that one before.

So there we were, a group of super post-humans whose bodies and brains were forever young, beautiful, and invincible, each of us an Immortal Me Bot that had been designed and rebuilt over the byrlennia to last forever, or until the universe eventually crunched in on itself, whichever came first. Of course, anyone could turn themselves off at any time, although most of us who had lasted this long without offing ourselves weren't likely to take that route, even if every century or so you heard about someone flipping that switch.

Admittedly, for countless myrlennia this whole Immortality Me Bot thing was seriously fun, what with all the genetic and quantum upgrades. Most of this technology eventually became biological, although we got a few key Humano-Machine (HM) augmentations, too. These were a little bit like the ancient apps that were once downloaded into smartphones, except that HM apps in the future were downloaded directly into our brain's organo-circuitry, brains that over the eons have become so engineered and reengineered that it is impossible to know where the carbon leaves off and the silicon begins. Or where the original gray matter ends and the augmented bio-organoids and dendrite-axon plug 'n' plays begin.

Far off in the future the Googleapplezon app store carries over seventeen quadrillion neuro-apps. Which was way too many, even for one million times Einstein smart brains that were already feeling like there was way too much going on inside noggins brimming with all those upgrades. But don't get us wrong. Our lives in the future really have been pretty fabulous compared to pre-ancient times. Yet there is also a slipperiness about who we really are when we can be anybody or anything we want, and when we travel hither and yon to any point in time between the Big Bang and the end of the universe (see "God Bot"). This has left us wondering for what seems like forever if we humans in the future were facing the ultimate existential challenge to our survival—what future philosophers called "terminal ennui."

We should hasten to add that not everyone in the future was bored. A subgroup of superhumans still woke up each morning excited to be living basically forever, and never ran out of things to do. This tiny but very vocal group of immortals was led by one of the oldest people in the universe, a human named Marc Hodosh. Tall, dark-haired, and almost always dressed in black, Hodosh had an obsession with ending death and making people immortal that began billions of years ago, back when humans actually lived for only a few decades, and then, gulp, actually *died*.

"I think everyone should not have to die," said Hodosh way, way back in 2018. A few years earlier he had cofounded TEDMED (the medical and health version of TED); he also started companies and produced television shows and events, often with a longevity theme. "To me in the future it should be a tragedy on a national, worldwide scale if someone dies. I don't think there is anything after death, no heaven, no reincarnation, therefore we should want to protect life to the extreme.

"Time," continued Hodosh, "is fictional; something man-made that represents a state of being. Getting older, for example, is not about time itself but about your state of health, mind, and your overall state of being. If I'm engaged with life, love friends and family, doing interesting things—that's a state of being, too. Protecting a healthy 'state of being' is more important than focusing on our perspective of time."

Hodosh was asked about the possibility of immortality being kind of boring. "The idea we're going to be bored is silly to me," he said. "Are you more bored at eighty than you are at forty? I don't think so. I think if anything we want to gather even more knowledge." According to the Hodosh creed, there would always be so much to do and experience and feel that ennui would never be a problem. "Let's say in a billion years we have explored eighty percent of the universe," he said, "but there is still twenty percent left—so let's go check it out! You can learn to play every instrument; you will want to have every career. You will want to be a professor or an astronaut or a lawyer or a doctor; you can be anything.

"I think we will want to be able to greatly increase knowledge, to store knowledge, and this will go on indefinitely. I think there is an unlimited number of things to do, and you can do them again, and you can try variations of them because the world will keep evolving. It's not just going to stop. So many people will want to be part of that evolution, to create the next step. That's why we love the next iPhone. I think there will always be the next iPhone, even in the future."

He waved away concerns, like the fact that he was then in his forties and had no idea what it was like to be eighty years old in an unenhanced body. Or how humanity back then would feed and provide happiness to all the additional humans who would still be around if everyone stopped dying. Or that some people in this long-ago era didn't have the opportunity, or the bank account, to get their first advanced degree, like becoming a lawyer, never mind also becoming a doctor or an astronaut. "We'll figure all that out," he said. "We'll have the time." Then someone mentioned that the universe may end in a few billion years, which might mean that no one could actually live forever. Hodosh answered this by suggesting that post-human immortals living billions of years from the present would figure out how to cheat death even when the universe crunches or expands into nothingness or whatever it will do at the end.

But Marc Hodosh and his ilk were a small minority in our way-off future, as most people continued to feel a gathering universe-weariness. That's when a near-miracle occurred that no one had expected—the sudden appearance of two robots that actually had been dreamed up in pre-ancient times, only to be forgotten.

The first of these bots to rescue us from rising ennui was Purpose Bot, conceived by the pre-ancient Harvard longevity scientist David Sinclair. An Aussie known for his unabashed advocacy of living a very long time before that notion became mainstream or technically feasible, Sinclair was asked in 2018 what kind of robot he'd like to meet in the future. Surprisingly, he didn't say a longevity bot. Instead he said that he'd like to meet a purpose bot, a machine or AI program that could help people in the future find purpose and meaning in their very long lives once immortality had been achieved, something Sinclair assumed would happen sooner or later.

"I want a bot or bots that allow my family and friends to exist with me, and me with them far into the future," said Sinclair, whose narrow oval face and shy smile hid a temperament that was dogged and unrelenting in the pursuit of his quest to understand the science of aging and antiaging. "In the future, even if you can learn all about everything by talking to all these supremely brilliant bots, once you've learned all that, then what?" he asked, anticipating our ennui in the future. "Then it's just loneliness. In fact, the knowledge of why we exist, why we're here, where the universe came from, that may end up being far more disappointing than we would like. Because as we learn more about our place in the universe, the more we may realize how pointless it is. There probably is no meaning"—which is exactly how we actually feel in the future.

"This is why you need a purpose bot," said Sinclair. "Which is a way of saying that you need access to what really matters, to what gives life purpose and meaning, which is your friends and family. I do want to be able to do things beyond what we can currently do. Visit new stars, visit new solar systems. But I want to share this with my friends, my family—hopefully all of humanity. And we can live for millions of years, which will allow us to travel around the galaxy, never worrying about getting sick, and never worrying about dying, unless your spaceship crashed into a sun or something. That would be the future that I would like to live in."

Which is basically the future we got millions of years later.

"But it has to have a human touch," said Sinclair, "or this great long life isn't worth living. Actually, it doesn't have to be human. It could be a social interaction, but it has to be a long social interaction," where we are connected with people being people: curious, asking questions, caring about us, and we about them. "I want to explore and learn about the universe and what life is like in the future, but only surrounded by people that I can talk to about it.

"We also need the bots to give us purpose," he said. "It's not just one long vacation. In fact, a group of bots, whether they be colleagues or family, would help extend the human knowledge. The worst world would be if we had already discovered everything. Then what are we going to do? Sit around and drink cocktails? This is a big concern, that once we've solved everything, what will we need to learn? We can't just have hobbies and vacations for the rest of eternity. The reason that we enjoy life now is the social interactions and the discovery of new things. This is why we will want to surround ourselves with people that we love." Purpose Bot would enable this no matter where your friends or family members are in the universe, or what form they have taken—digital, or whatever.

"This bot would also ask us questions and keep us in line," said Sinclair, "because we all have people in our lives that challenge us, the naysayers, and that's a healthy thing, we need that. Otherwise we'd just become godlike figures in our own minds. I like to surround myself with people who are smarter, and with critics. That's the only way you can solve a lot of problems and provide yourself with mental stimulation that will keep you from being bored."

Back in 2018, Marc Hodosh agreed with Sinclair about not being bored, but he didn't think that people in the future would need a purpose bot. "Purpose comes from within," he said. "It's part of you."

Right. Except that most people in the way distant future weren't as excited as Hodosh about being an Immortal Me Bot. They struggled to keep finding purpose from deep within as the myrlennia and byrlennia rolled past. For these folks, Purpose Bot was a godsend, although there was still something missing in the future even when Purpose Bot restored meaning in their lives. That's where the second robot that saved us from terminal ennui came in.

This was Mom Bot, a remarkable invention that was first conceived in the same pre-ancient era as Purpose Bot, by a physician and investor named Joon Yun. "This is a bot whose intentions will be to help you be successful as a person," said Yun. "Mom Bot would support you with whatever is in your best interest with a kind of unconditional love. This is not a bot that solves problems; it's a bot that helps you with intent. It serves your truth, rather than serving someone else or serving itself. Bots that serve themselves [and the companies that build them] are the bots I fear.

"The bot that I want to create," he continued, "is the one with the intent to do the exact opposite" of a problem-solving bot that is working just for itself or a company. "Any machine that can self-learn is going to be sitting on top of us. It will take away our freedom," which we will have to be very careful about preserving in the future. "The solution to this will be the Mom Bot that has unconditional love for us, because they will serve us."

But don't some moms have their own issues and agendas that can mess up their kids?

"Moms can be tough," said Yun. "There can be issues with your mom. But the Mom Bot I want has this unconditional-love system, without all the things that can undermine the mom/child relationship. Because unconditional love can also be tough love. Sometimes you're your own worst enemy and it actually is in your best interest to have somebody come down pretty hard on you, but to help you, not as part of their own agenda."

Incredibly, what Joon Yun described back in the pre-ancient era is exactly what we got in the distant future with our Mom Bots. Because really, who wouldn't want a bot that showed us unconditional love and encouraged us to have a positive intent but also could be tough and didn't hesitate to tell us hard truths? Combined with advanced Purpose Bot tech, Mom Bots for a long time helped us in the very distant future to hold ennui at bay.

When Marc Hodosh back in 2018 heard about the concept of a mom bot, he rolled his eyes. "I don't think you'll need these support bots," he said. "Of course, we all want to feel like we have a place in the universe, and we want the love of friends and to have a purpose. But we don't need to get this from a machine. We get it from ourselves."

Inevitably, billions of years in the future, even our Purpose and Mom Bots couldn't fend off a gradual return of the dreaded ennui, which began to overtake us once again as the end of the universe approached—which immortal humanity had been anticipating for a very long time, wondering what would happen. But now that it was finally upon us, most of us could barely muster the energy even to yawn.

Not so with Marc Hodosh and his followers. When the moment came at last for the universe to end, Hodosh and his followers weren't lollygagging around, trying to find purpose and love. No, Hodosh and his posse were working hard on a singular task that took them billions of years: figuring out how to cheat death yet again and survive the collapse of the universe into a small cluster of matter known as the big crunch. "I'd want to survive the end of the universe," Hodosh had said billions of years earlier, back in 2018. "That way I get to see what happens next, after the end of the universe."

True to his word, Hodosh and his scientists just before the end of the universe announced they had determined with a 99.9999 percent certainty that the universe after crunching would immediately recycle and launch a new universe with another Big Bang, a process of explosion and implosion and back again that Hodosh and his scientists claimed had been going on forever, with an infinite number of possible universes resulting.

That might sound like a damn impressive finding to people in the distant past, but to us living at the end of the universe it was ho-hum, even if we tried really hard to get excited by turning to our Purpose and Mom Bots that previously had been so helpful in fending off terminal ennui. But not even a look of daggers in the eyes of our Mom Bots set to "tough love" mode could convince us to give a damn about one of the greatest scientific achievements ever.

Then Hodosh and his followers topped that by claiming they had invented a new technology that would allow them to survive the end of the universe. They unveiled a device that they said would create a memory loop made of subatomic particles that would lodge themselves inside the final small cluster of molecules that the universe was going to collapse into. This would allow the essence and memories of Hodosh and his followers to be reconstituted in the new universe as it big-banged and expanded. They weren't sure exactly how this would manifest in the new universe, except that they were certain they would exist and know who they were and that they would be able to interact with one another.

Eventually, Hodosh and his scientists said, life in the new universe would be sparked on an Earth-like planet, where evolution would produce creatures resembling humans. In fact, they claimed that this process had been happening through countless Big Bangs and contractions forever, with a handful of people from past iterations of universes surviving to become wise men and gurus and others we describe as "old souls." The difference this time, they said, was that the new versions of themselves in the universe to come would remember that they were Marc Hodosh and his friends.

Right, we said, not interested enough to even try to follow all this even as the moment came for our universe to end and we said adieu to our buddies and loved ones—and of course to our Mom Bots. Once again bored out of our minds, we merely shrugged as the universe began to rend itself asunder and atoms smashed into one another, although we have to admit that most of us did grip our Mom Bot's hand a wee bit tighter than we ever had before.

As far as we knew, just before our phenomenally long lives as Immortal Me Bots ended with this massive, cataclysmic contraction of the universe back into a single cluster of molecules floating in space, Marc Hodosh and his followers did make it to the other side. What they found we could never know, which we'd like to say aroused at least a tiny glimmer of curiosity in us as our molecules dissolved and joined the massive swirl of matter falling in on itself. But truth be told, we weren't the least bit curious as we faded and slipped away into a vast and delicious nothingness.

EPILOGUE
AFTER THE ROBOTS ARRIVED

In the future, we will all recall when we first really talked to robots and they talked back to us. Admittedly, there will be times when we will have no idea what they are saying, and vice versa. Yet we fervently hope that we at least come to an understanding with the remarkable machines that we are building with our post-ape-savanna brains—that they work for us, and not the other way around.

As we look back on the ERE (Early Robot Era), we hopefully will see that humanity managed to muddle along as the years went by, just as people always have—with occasional flashes of brilliance and epiphanies of wonder: when Coffee Delivery Bot finally placed that warm cup of joe in our eager hands or when our Daemon Bot kept us from falling off a red cliff on Mars. And there was that magical moment when God Bot revealed the secrets of the universe as we dashed hither and yon in those sleek and beautiful Timeliner ships powered by multidimensional-neural-uncertainty fields, taking us from the Big Bang to the implosion and crunch at the end of the universe and back again.

We recall the darker moments, too. Including those where we were really kind of stupid, letting humanity's worst impulses get passed on to the core coding in our machines. In these futures, how could we forget the dumb-ass attributes that have always caused us problems, ranging from the merely annoying, like when Journalism Bot wrote in clichés, to the staggeringly tragic, like when Warrior Bot nearly killed us all (except that was a movie, right?). In between, as the future unfurled, we had so much fun with our Teddy Bots, Tourist (Evolution) Bots, and Wear-Bots. And, of course, we loved our Intimacy Bots, although as your Intimacy Bot would not hesitate to tell you, sex with a manta ray should never be a substitute for intimacy and sex with real humans!

We also got to watch as our robots and AI systems went through their machine-learning protocols, starting out, let's face it, a little bit stupid (but hey, Alexa, we still love you, at least we think we do). This came even as we experienced our own "human-learning" curve to make us smarter vis-à-vis our machines. Like when we decided to stop 👍'ing Facebook Bot until it ceased and desisted already with the troll-hackers from Russia and North Korea, and with the uncurated social media machinations of power-mad politicians, nutjobs, and unscrupulous advertisers that like real trolls should stay under slimy rocks and old, abandoned, moss-covered bridges that only an idiot would cross in the dark of night.

One of the key things we learned about our machines was that we didn't have to build every damn robot and AI system just because it was cool, and because we could, and because it might make a small number of people lots of money in the short term. The

WAMBS taught us that. So did Sunny Bates when we humans rather foolishly allowed the robots in one future scenario to take every job on the planet. And remember when we thought we loved our Doc Bots more than human docs? Okay, some of us really did bond with those sleek, shiny Doc Bot slabs with the face of George Clooney as Dr. Doug Ross. But for the parts of medicine where we craved empathy and wanted to tell our stories' beginning, middle, and end, we really did like having a human touch. We also loved having Thriller Bot working with great human writers to provide at least a few fast-paced narratives that were 100 percent great (not 90 percent) while we waited for our appointments with our human physicians, who still, for some reason, even in the future, couldn't stay on time.

And then there was the time that we came perilously close to inadvertently allowing Warrior Bot to think it was playing a game of Go or Global Thermonuclear War, which it was programmed to win at all costs—a prospect that remains seriously terrifying in both the present and the future with the proliferation of swarming, hummingbird-size drones, airplane-mounted lasers, bioengineered pathogen-bombs, cyber-attack bugs, squishy chemical bots that can deliver bombs through keyholes, and modern-day WOPR attack-and-kill systems that we hope will never, ever, *ever* be automated. As Major General Bob Latiff suggested back in 2018, the possibility that things could turn out very badly with Warrior Bot is not just a plotline in a scary sci-fi film.

Hopefully, we'll remember that it's really not about the robots, even if certain robot politicians might try to seduce us with the idea that AI politicians will be paragons of truth, decency, and all that. In the future, we know better—or at least we think we know better—understanding that AI presidents, CEOs, Supreme Court justices, and even baristas are only as good and truthful as we make them.

As Dean Kamen said, technology is all about amplifying things—good or bad, fun or not so fun, and, with beer, rich and foamy or not. It's kind of up to us, particularly if we reach the point where we will want to go *digitalis* or *syntheticis* and push ourselves into a paradigm of self-evolution that takes us way beyond being human. Realizing, of course, that being forced to decide between going digital and going synthetic is actually a silly and false choice since if this stuff ever really became possible, we are likely to choose a combo of the two in the ultimate Human + Machine manifestation. Unless, like Juan Enriquez, we are so desperate to go to Proxima Centauri b even when we can't go faster than 186,000 miles per second that he isn't willing to wait until God Bot appears to teach us how to build Timeliners, which will be able to whisk us anywhere we want in the universe, and in any dimension, in no time at all.

But don't worry. To pass the time before God Bot arrives, we've got Opti-Brain Bots to keep our minds healthy and Purpose Bots and Mom Bots to keep us focused and inspired about what matters, even as we scratch our heads and wonder if we are living in some version of the Matrix.

As the future unfurls, we'll keep talking to robots, and they to us. After all, talking to them is like talking to ourselves. Really, it's not that different from when our ancestors chatted with and prayed to their gods of love, wisdom, and war, since they were actually talking about aspects of humanity that we love and hate about ourselves, often at the same time. This begs the question: Even billions of years from now, will our machines help us understand our contradictions, use our contradictions to destroy us, or just find the whole human experience weird and amusing? This assumes that we make it that far and that Alexa and Google Home make the huge leap from being there to turn up the AC and play the greatest hits from the Beastie Boys to taking us at Warp 6 to Star Base 62.

But hey, people of the present, we've already told you enough about your possible futures.

Now it's up to you to be at least as smart as your phone or your toaster, and to do the right thing, okay?

ACKNOWLEDGMENTS

I am grateful to the many humans who helped enormously in making *Talking to Robots* happen. This starts with Stephanie Losee, who encouraged my nutty idea to write a book about robots that's really about humans told from the future. The other person who didn't laugh was my agent Mitch Hoffman, who by good fortune (for me) decided to switch from being a highly successful editor to agenting; and who helped me develop the proposal and land it at Dutton, where my editor, John Parsley, has been patient and spot-on in helping shape this narrative and guide me through the publishing process. As always, thanks to my agent in the UK, Eugenie Furniss, for everything over the years—and for also encouraging me early in the idea phase. Also at Dutton, thanks to Cassidy Sachs, Emily Canders, Kayleigh George, and Andrea Monagle.

Deep appreciation goes out to the subjects of this book, who allowed themselves to be incorporated into stories set in the present and future—with a special shout-out to Rodrigo Martinez, my co-conspirator and brother in arms. Special thanks to those who read and, in some cases, reread drafts of the manuscript and provided sage commentaries, including Robyn Heister, Iya Khalil, Marc Hodosh, Marjolein Brugman, Minou De Gruyter Palandjian, and Susan Feeny. And to Mom and Dad for encouraging me. Thanks also to Minou, Robert Green, Sally McNagny, and Lenny Barshack, for providing some very special places to write. Also, to my friends who put up with me talking too much about robots, including the aforementioned friends and readers, plus Gillian Sandler, Kristin Hamann, Jamie Heywood, and Zuberosa Marcos. Finally, thanks to the entire *mucha ropa* tribe who supported and inspired me, fostering the project in Tulum and beyond!

ABOUT THE AUTHOR

David Ewing Duncan is an award-winning science journalist in print, television, and radio, and the bestselling author of nine books published in twenty-one languages. He is the curator of Arc Fusion, a recent columnist for *The Daily Beast*, and a frequent contributor to *Wired, Vanity Fair, MIT Technology Review, The New York Times, The Atlantic,* and others. He has also written for *Fortune, National Geographic, Discover, Life, Outside,* and *Newsweek*. He is a former commentator for NPR's *Morning Edition*, and a special correspondent and producer for ABC's *Nightline*. His book *Calendar* was a bestseller in fourteen countries; he also wrote the bestseller *Experimental Man*. David's work has won numerous awards, including Magazine Story of the Year from AAAS. David lives in Boston and San Francisco, where he is a member of the Grotto, a San Francisco writers' co-op.